Mechanical Sound

INSIDE TECHNOLOGY
edited by Wiebe E. Bijker, W. Bernard Carlson, and Trevor Pinch

A list of books in the series appears on page 315.

MECHANICAL SOUND

TECHNOLOGY, CULTURE, AND PUBLIC PROBLEMS OF NOISE IN THE TWENTIETH CENTURY

KARIN BIJSTERVELD

THE MIT PRESS
CAMBRIDGE, MASSACHUSETTS
LONDON, ENGLAND

This book was set in Bembo by Graphic Composition, Inc., Bogart, Georgia, using InDesign CS2.

Library of Congress Cataloging-in-Publication Data

Bijsterveld, Karin, 1961–
    Mechanical sound : technology, culture, and public problems of noise in the twentieth century / Karin Bijsterveld
        p. cm. — (Inside technology)
    Includes bibliographical references and index.
    ISBN 978-0-262-02639-0 (hardcover : alk. paper)—978-0-262-53423-9 (paperback)
    1. Noise pollution. 2. Sound—Social aspects—History—20th century. 3. Machinery—Noise. 4. Noise control. 5. Noise music. I. Title.
TD892.B548 2008
    620.2—dc22   2007037600

do not close the windows
it will enter soon
the very quiet

[sluit de ramen niet
straks zal het binnenkomen
het allerstilste]

—J.C. van Schagen, *Ik ga maar en ben* (Amsterdam: G.A. van Oorschot, 1972), p. 245.

CONTENTS

CONTENTS

One Sunday afternoon in 1994, I sat down in my garden and began thinking about my future research. Wiebe Bijker, head of the Department of Technology and Society Studies at Maastricht University, the Netherlands, had just offered me a job. His only condition was that I would focus my research on technology studies and come up with an appropriate topic. I had difficulties concentrating, though. I blamed the roar of aircraft, lawnmowers, and radios around me for disturbing my train of thought and I decided to go inside. On entering the house, I suddenly realized what the topic should be: noise! Wasn't it related to technology and a key phenomenon of modern society? And wasn't noise the perfect chance to combine my interest in sound and music with technology studies?

At that moment, I did not know that my blaming noise for interrupting my contemplation was part of a long tradition of complaint. Nor did I know that the people who had studied noise were, at times, like me, deeply involved in music. I would soon learn about these similarities, however. The many occasions for identification with those writing in regret or praise of the sounds around them have contributed to my great pleasure in doing the research for this book. Equally significant to this pleasure have been my visits to the libraries and archives that stored sources on the history of sound and noise in Western society. I will never forget my stay at the archives of the World Soundscape Project in

Vancouver, Canada. The room with the project's files was close to a composition studio where people were experimenting with the intriguing sound of a spinning wheel—a sound that I still hear when recalling that memory.

Ever since I started my research, and even in the years when a history of noise was just a research plan, colleagues within and outside of Maastricht have been extremely helpful in pointing out newspaper items, papers, books, reports, stories, poems, web sites, and conferences related to noise. Even now, not a week passes without a colleague's note in my pigeon hole or my email inbox about something relevant to my research. I would therefore like to begin my acknowledgments with simply thanking *all* my colleagues, including former ones, at the Faculty of Arts and Social Sciences at Maastricht University for being so helpful, most notably Jos Perry and Jack Post, who have been especially active in keeping me posted on sound and noise. Friends like Helma Erkelens, Maloeke de Groot, Dorien van Rheenen, Annet Perry-Schoot Uiterkamp, and Jeroen Winkels also came up with documents on noise once in a while.

Moreover, many of my Dutch as well as foreign colleagues have commented on the proposals, papers, articles, and chapters that I wrote along the way, as editors of journals, special issues, and books, experts in acoustics, or in other roles. I am deeply indebted to their critical remarks on and interest in my work. These commentors include Jan Baetens, Wiebe Bijker, Hans-Joachim Braun, Christian Broër, Michael Bull, Howard Cattermole, David Edge, Sven Dierig, José van Dijck, Ernst Homburg, Frank Huisman, Arnold Labrie, Jens Lachmund, J. Andrew Mendelsohn, Kathryn Olesko, Harry Oosterhuis, Joy Parr, Peter Peters, Trevor Pinch, Jan de Roder, Fort de Roo, Pieter Jan Stallen, John Staudenmaier, Manuel Stoffers, Emily Thompson, Rienk Vermij, Ginette Verstraete, Gerard de Vries, Jo Wachelder, Rein de Wilde, and many anonymous referees.

Yvette Bartholomée, Marten Schulp, and Ragna Zeiss assisted me, as student assistants or trainees, in finding relevant sources. Siegfried Böhm, Ivo Blanken, Hans Cauberg, Jef van Dongen, Amanda Engineer, Michael Fahres, A. C. Geerlings, Bert Hogenkamp, Ronald de Jong, Marnix Koolhaas, Jan Kuiper, Chris Leonards, Karljosef Kreter, Marianka Louwers, Hugo Paulissen,

Hillel Schwartz, Floris van Tol, Barry Truax, Lia Verhaar, Ludger Visse, Michael Voegele, Robert Vrakking, Peter Wakeham, Valerie Weedon, and Hildegard Westerkamp provided me with access to archives, libraries, catalogues or electronic sources. Ruth Benschop, Ton Brouwer, Geert Somsen, and Joke Spruyt corrected the English of single papers, while Margaret Meredith edited the full manuscript attentively, and Margy Avery advised me on how to transform the manuscript into a real book. I sincerely thank all of them for being so involved. I am also grateful to the Faculty of Arts and Social Sciences of Maastricht University for funding part of the editing of the book, and to NWO for providing both a substitution grant and a language correction grant.

A grant awarded by the Dutch National Science Foundation (NWO), for the academic year of 2000–2001, allowed me to concentrate on noise, to write several articles, and to write the first draft of the book. Portions of chapters 2, 3, 4, 5, and 6 have been derived from the following articles: Karin Bijsterveld, "The Diabolical Symphony of the Mechanical Age: Technology and Symbolism of Sound in European and North American Noise Abatement Campaigns, 1900–40," *Social Studies of Science* 31, no. 1 (2001): 37–70; Karin Bijsterveld, "A Servile Imitation: Disputes about Machines in Music, 1910–1930," in *I Sing the Body Electric: Music and Technology in the 20th Century,* ed. Hans-Joachim Braun (Baltimore: Johns Hopkins University, 2002); Trevor Pinch and Karin Bijsterveld, "Should One Applaud? Breaches and Boundaries in the Reception of New Technology in Music," *Technology and Culture* 44, no. 3 (2003): 536–559; Karin Bijsterveld, "The City of Din: Decibels, Noise and Neighbors in the Netherlands, 1910–1980," *Osiris* 18 (2003): 173–193; K. Bijsterveld, "Listening to Machines: Industrial Noise, Hearing Loss and the Cultural Meaning of Sound," *Interdisciplinary Science Reviews* 31, no. 4 (2006): 323–337. I thank each of the publishers (Sage Publications, Johns Hopkins University Press, University of Chicago Press, and Maney Publishing, respectively) for permission to reproduce relevant parts of these articles.

Once fully concentrated on research and writing, I tend to hear hardly anything anymore. Yet my daughter Sarah reminded me to keep listening. I expect her to enter the iPod mode soon, but when younger, her positive attitude inspired her to point out the "nice" and "funny" sounds surrounding us, such as those of our bicycles. Rein de Wilde, my partner and colleague, has been, as always, enormously important for discussing lines of arguments and for making research travels feel like holidays. What's more, he e-mailed the excerpt from Van Schagen's poem in 1994, now this book's epigraph, which stimulated both my research and my life at large.

Listening to Technology

A Persistent Issue

In 1875, the British hygienist Sir Benjamin Ward Richardson described an imaginary city of health named *Hygeia*. "The streets of our city, though sufficiently filled with busy people," he said, "are comparatively silent." Beneath each of the main boulevards "is a subway, a railway along which the heavy traffic of the city is carried on. . . . The streets of the city are paved throughout in the same material. As yet wood pavement set in asphalt has been found the best. It is noiseless, cleanly, and durable. . . . The subways relieve the heavy traffic, and the factories are all at short distances from the town, except those in which the work that is carried on is silent and free from nuisance" (Richardson 1875: 950–951). Needless to say, no such city existed at that time, and still today there is no city that is so silent that the discussion of noise has disappeared from public life—even though we have asphalt, subways, and industrial parks.

Complaints about noise have been recorded throughout history. Yet beginning in the last quarter of the nineteenth century, such complaints became increasingly focused on new technologies: on the sounds of factories, trains, steam tramways, automobiles, and gramophones. In essays and pamphlets, lively descriptions of all kinds of noise were given. By the early 1900s, antinoise leagues had been formed all over Western Europe and North America, organizing antinoise campaigns, antinoise conferences, antinoise exhibitions, and "silence weeks." The ensuing public debate about noise has never died down. What made noise such a persistent issue on the public agenda? How did sounds become the subject matter of public problems, that is, of problems pushed into arenas of public action? And what rendered these problems of noise so hard to tackle?

Today's popular publications, policy documents, and academic reflections on noise provide three answers to the question of why noise is such a persistent problem. The most common response relates to economic and population growth. This growth has led to a world inhabited by ever more people who are ever more mobile and possess ever more noisy equipment. The sheer increase in the quantity of sound sources has left all the improvements in noise control inaudible, as it were. A second response considers the specific characteristics of hearing a major factor. Unlike our eyes, so this argument begins, we cannot close our ears. We continuously need our ears for information and communication, so sound, even though inherently transient, is always around. Hearing has a highly subjective side to it: sounds that annoy some people are music to the ears of others. Since noise is widely defined as "unwanted sound," the subjectivity inherent in this definition complicates legal intervention when rival definitions of noise arise. Finally, there is the so-called visual regime of Western culture: in the West's hierarchy of the senses, the eye dominates the ear. This makes sound into a neglected issue. Even worse, our culture is deadly afraid of silence and of the passiveness associated with the absence of sound. Those who try to explain the difficulty of managing public problems of noise, then, tend to invoke arguments that refer to our culture's interest in, if not its obsession with, economic growth, to the innate characteristics of hearing, and to the apparent sensory priorities of our culture.

To be sure, many of these answers make sense. They refer to basic dilemmas and features of Western society. Yet they suggest a degree of continuity in explanations for the difficulties of dealing with noise over time that conflicts with even a superficial encounter with our sonic past. What, for one, are we to make of the observation that in the 1930s the problem of noise was predominantly phrased as a "honking problem," whereas this problem has now virtually vanished from the public agenda even though we still highly value automobiles as a means of transportation? And what do we do with the finding that the subjectivity of sound perception we now believe in had a quite different status in 1875 and may not have hampered the approach to noise in the same way as it does today? Such changes in the definition of the public problem of noise over time imply a succession of *different* public problems of noise and make ahistorical explanations for the persistence of noise on the public agenda less powerful.

A comparison with stench may further enhance our awareness of the significance of the relation between the definitions of problems and their context. In the nineteenth century, the public nuisance of stench was a problem of sense and sensibility that came to be dealt with in a far more comprehensive way than noise has ever been. The identification of stench as the seed of contagious and dangerous disease by an elite that managed to intervene deeply in private households led to its confinement. It is no coincidence that the strategies Richardson proposed to abate city noise were perfectly analogous to the ones he thought fit for getting rid of "the foul sight and smell of unwholesome garbage" (Richardson 1875: 950). The subways for heavy traffic resembled the sewage-subways for washing away mud, and putting industry at a distance was like compiling trash at the margins of the city. Yet the contexts for noise abatement happened to be rather distinct from the ones in which stench came to be tackled.

In order to be tackled thoroughly, public problems need convincing drama, robust definitions, and empowering coalitions, both within and, increasingly, across nation-states. The problem of noise, however, has never, or only temporarily, met these requirements. Instead, this book argues that the rise of new machine sounds and the process of stacking various forms of noise legislation on one another over time created a *paradox of control*. Experts and politicians increasingly promised to control noise by measuring and maximizing sound levels. Yet they defined some problems, such as neighborly noise, as difficult to capture in quantitative terms, and left it up to citizens to talk their neighbors into tranquil behavior, while wrapping other issues, such as aircraft noise, in formulas beyond citizens' reach. Citizens have thereby been made responsible for dealing with the most slippery forms of noise abatement and distanced from the most tangible ones. This has not exactly helped to wipe noise from the public agenda. The *spatial* character of many of the interventions has similarly contributed to the persistence of public problems of noise. It is remarkable that sound, crossing the borders between neighbors, cities, and nations so easily, has often been handled spatially, for instance, by imposing zones, canalizing traffic, and drawing noise maps. We have been trying to create islands of silence, yet have left a sea of sound to be fiercely discussed.

This book explains how we ended up like this. It focuses on four crucial episodes in the Western history of noise between the late nineteenth and the late twentieth century: public discussions of industrial noise, of city traffic noise, of neighborly noise of gramophones and radios, and of aircraft noise. A fifth chapter highlights the celebration of noise in the avant-garde music of the interwar period, and thus serves as a *counterpoint* to the other chapters. It both illustrates how such reverence embodied the positive connotations of mechanical sound that antinoise activists had to cope with, and shows how the introduction of machines in music re-enacted the issue of who was to control sound. The remaining chapters explore the decades immediately succeeding the rise of the public debate over the roar of new, or recently ubiquitous, machines. In doing so, this book centers on society's struggle and occasional success with controlling mechanical sounds. It also underscores how the strategies for solving earlier noise problems—embedded in law, scholarship, scientific instruments, and techniques—recurred in and often structured the approaches to newer ones, which at times created new problems. How can we account for such continuities? And what can we learn from the fate of former noise abatement strategies when thinking about contemporary problems of noise? But let me first unpack and underpin some of the statements above.

"WE CAN'T STAND IT ANYMORE": PUBLIC PROBLEMS OF NOISE

Noise is a popular topic among today's Dutch newspaper columnists. One of them mocks men who come home from work at five o'clock only to mow the lawn and trim the hedge with as many noisy machines as possible (Mulder 2000).

Another columnist regrets that even in the most expensive hotels one tends to hear the neighbors, except if one has the air conditioning on (Lagendijk 1998). Others lament over the boom box of some carpenters at work a few homes down the block, over the noise created by the seemingly eternal remodeling project that is going on next door, or over the everyday noise of the buses and trucks down the street. The sound of aircraft traffic, mobile phones, restaurant music, and the steam explosions of cappuccino machines—it is just too much (Vreeken 2000; van't Hek 2000; van der Laan 1999; Abrahams 2000; Ritsema 1995; Pessers 1997). People cannot stand it anymore: the noise of compressors, the radio at work, the music in the supermarket—the absence of silence (van Delft 1995; Campert 1997; Doves 2004; van Renssen 2002). Many columnists emphasize the omnipresent sounds of today's technology: the whirring of the video tape, the hiss of the television standing by, the hum of the refrigerator, the buzz of the electricity gauge, the click of the heating pipe, and the roar of the fan (Blankesteijn 1998).

The subject of noise is so common in these occasional pieces in Dutch newspapers and magazines that one can speak of a distinct genre. The columnists respond to the politics of the day, or aim to raise public consciousness. In many ways, perhaps, their choice to discuss this topic is an act of distinguishing themselves from the ordinary people. *We,* they seem to be saying, are not like all those men who come home from work and have nothing else to do than start making noise; we are not the neighbors who endlessly remodel their homes; nor are we the owners of boom boxes. Others may not notice, they suggest, but we certainly *hear* all the sounds that others somehow feel the need to generate. Tellingly, many of these columnists discuss the topic of noise with a sense of humor, a touch of self-irony, or just enough feeling for rhythm to allow the reader who does not sympathize with their complaint to at least admire their style of complaining.

If the sounds that prompted these Dutch columnists' reflections are rather new, complaints about noise in the popular press are not. The World Soundscape Project, a research project housed at Simon Fraser University (Vancouver, Canada) directed toward documenting a history of the world's changing sonic environment, has examined how often "noise" has shown up in the news in the past in a study based on a survey of sixty-five magazines published in North America between 1892 and 1974. Until 1926 there were usually fewer than five entries on noise each year. The titles of many of these early articles refer to city and street noise. In the second half of the twenties the number of articles featuring noise began to increase, peaking at twenty-one in 1930. After 1930, the annual number of popular articles on noise never exceeded thirteen, until 1968, when a second peak occurred. In 1974, the last year of the study, the situation is back to "normal." [1] Most likely, the total number of popular articles on noise has even been higher. A campaign against city noise undertaken by the New York Noise Abatement Commission between 1929 and 1930 was accompanied by at least "130 newspaper articles throughout the United States and Europe commenting on this project" (Dembe 1996: 201). In 1962, "one day's press cuttings" by the British Noise Abatement Society produced 151 items. Among the headings focusing on one type of noise, aircraft noise was the most dominant (Some Headlines 1962: 22–24).

There is clearly a correlation between articles about noise and particular antinoise campaigns or activities. This suggests that heightened noise abatement activity indeed fuels public debates about the issue. However, it would be wrong to suggest that there is a one-to-one relationship between the level of attention given to noise in newspapers and magazines and citizens' complaints about the nuisance of everyday noise. The findings of a study on self-reported nuisance conducted by the Dutch National Data Agency in 1997 are illustrative. It found that 27 percent of respondents said they were disturbed by traffic noise, 21 percent by neighborly noise, 19 percent by aircraft noise, and 11 percent by industrial noise (de Jong et al. 2000: 66). The same year, the Dutch Noise Abatement Foundation published a report on the number of times the issue of noise appeared in Dutch newspapers between May and October 1997. Aircraft noise topped the list with 2,663 news items, whereas road traffic was mentioned in only 293 items, neighborly noise in 240, leisure-related noise in 237, noise related to industry in 231, and noise related to train traffic in 193 (Aantal 1997: 3). These data suggest that the nature and level of attention paid to the issue in the press does not automatically correspond with how the average individual evaluates particular noise problems. There are over ten times as many publications on aircraft noise than on the noise of neighbors, but the inconveniences caused by the people next door are higher on the list of nuisances reported by the people interviewed.

Such findings make clear why *public* problems should be distinguished from *private* ones. In his study on the culture of public problems, Joseph Gusfield claimed that not all problems "necessarily become public ones" in the sense that they "become matters of conflict or controversy in the arenas of public action" (Gusfield 1981: 5). As an example, he refers to people's disappointment in friendships. Even if feelings of disappointment may be very painful on an individual level, so far no public agency has been set up to solve this problem. Perceptions of these sorts of problems can change in the long run, however. Teasing individual kids at school has probably been a problem ever since schools have been around. Only recently, however, have Dutch schools been made formally responsible for the problem after the assistant secretary of education issued a regulation forcing schools to take action against it. In Gusfield's terminology, the Dutch assistant secretary is now one of the "owners" of the *public* problem of teasing, whereas the school is charged with solving it. The discrepancies in the hierarchy of noise problems—between the level of attention for them in the press and the ways in which people experience them—is an intriguing indication of how the character of problems changes when problems are transformed from private into public ones, or when they change from one public arena to another.

Public problems of noise are currently "owned" by hundreds of organizations, institutions, and industries created with the purpose of abating, regulating, or studying noise. Almost every country in Western Europe has at least one nationally operating noise abatement organization. These are, for example, the Ligue Française contre le Bruit (French Anti-Noise League), the Nederlandse Stichting Geluidshinder (Dutch Noise Nuisance Foundation), the British Noise Abatement Society, and the Deutscher Arbeitsring für Lärmbekämpfung (German Working Group for Noise Abatement). Most of these agencies were founded between the late 1950s and 1970s, but many had forerunners that date

back to the first decades of the twentieth century. Present-day European organizations are usually members of the Association Internationale Contre le Bruit
(International Association Against Noise), established in 1959 (Lehmann 1964:
11). Outside Europe, noise abatement organizations are found in many countries
including the United States, Canada, Israel, and Argentina. And often cities and
even neighborhoods have their own antinoise groups, such as those found in
Berlin, Washington, and New York (including the Bronx), to mention just a few
examples.

In addition to agencies that address different kinds of noise problems,
there are many that focus on particular types of sound. The most common
ones are those that fight the noise produced by a specific airport.[2] Others, such
as Pipedown International in the United Kingdom, deal with background
music. Violently sounding acronyms are quite common. BAM is the acronym
of the Dutch lobbying group against Muzak; BLAST, located in Santa Barbara,
California, stands for Ban Leafblowers and Save our Town, and HORN (Mt.
Tabor, New Jersey) for Halt Outrageous Railroad Noise.[3] And there has been
an annual World Noise Awareness Day since 1996. At this and other occasions,
organizations like the League for the Hard of Hearing frequently ask national
governments and international bodies to take public action against noise. Indeed,
countless governmental committees, ministerial departments, national health
councils, and standardization organizations have entered the realm of regulating noise, defining the noise problem and distributing responsibilities on their
own account. So have their international counterparts, including the European
Union and the World Health Organization. These agencies seek expert advice
from scholarly organizations for acoustics, noise control engineering, and audi-
ology, or from countless acoustic consultants.[4]

This overwhelming and still expanding network of initiatives for noise
abatement and control makes clear that noise is on the minds and in the hands of
many. Almost nowhere, however, has the problem of noise been removed from
the public agenda. This may explain the noisy names of some of the pressure
groups. What has gone wrong? Why is noise an enduring if not a permanent
public problem?

THE PUTATIVE DIFFICULTIES OF DEALING WITH NOISE

Today, the difficulty of tackling the public problem of noise is most commonly attributed to three causes: the economic prerogative of growth that conflicts with a quiet lifestyle, the subjectivity of hearing, and the intrinsic character of Western culture. In the first line of reasoning, noise involves a hard to solve problem because it results from a fundamental conflict between economic progress, population growth, and increasing mobility on the one hand, and concerns of public health and the environment on the other. As the journalists Peter Müller and Marcus von Schmude noted in the German weekly *Die Zeit,* there is much ado about noise-induced health problems, yet the millions of people "who complain about traffic noise take the car nonetheless (2001: 9)." One Dutch journalist portrayed these millions as victims. Their decent lives, she stressed, have been subordinated to the interests of the "supreme" transport industry, "the God of the twentieth century" (Pessers 1997). The basic opposition is the same: the public's well-being is being exchanged for mobility.

In 1996, the European *Green Paper on Future Noise Policy* made a similar claim in less accusatory prose and more detail. It stated that since 1970 "the noise from individual cars has been reduced by 85% . . . and the noise from lorries by 90%. . . . However data covering the past 15 years do not show significant improvements in exposure to environmental noise. . . . The growth and spread of traffic in space and time and the development of leisure activities and tourism have partly offset the technological improvements."[5] According to Egon Dietz, a staff member of the Dutch National Data Agency, the Dutch government spent 2.8 billion guilders on noise abatement between 1979 and 1993. During the same period, however, the percentage of citizens who complained about noise hardly decreased at all. The growing population, rising population density, increasing mobility, and the widespread possession of audio sets, Dietz stresses, have all contributed to the complexity of the noise problem (Dietz 1995: 20).

The second type of argument provides reasons for the enduring trouble of noise by referring to the characteristics of hearing. As the noise historian Hillel Schwartz has shown, the baseline to "the litany of twentieth-century antinoise polemics is the claim that human hearing is constant, involuntary, and nearly impossible to shut off" (2003: 487). These features explain, at least to many people writing about noise, the difficulty of dealing with noise problems. Our sense of hearing needs to function constantly because it provides us with crucial information. "It is easy to imagine how dangerous a completely silent car would be," an architect proclaimed in 1967. "What, in fact, we are combating is not so much noise as such . . . as its dual character. We are trying to abolish noises that are harmful to human beings, but not to get rid of all noises, since this would deprive man of a vital source of information" (Stramentov 1967: 8).

What's more, the perception of sound is now considered to be highly subjective. Psychologists argue that whether individuals are annoyed by a specific sound is not only dependent on the characteristics of that sound, such as its loudness, frequency, or periodicity; equally relevant are one's physiological sound sensitivity and compulsivity, as well as the social context and perceived control (Hell et al. 1993: 247). Dietz believes that this subjectivity of sound perception accounts for the persistency of discontent (1995: 21). Or as *Die Zeit* journalists put it, "Noise separates beergarden-friends from tranquility-lovers, techno-fans from visitors of chamber music concerts, churchgoers from late risers. . . . And that's why noise abatement is hard" (Müller and Von Schmude 2001: 9). The social scientist Ronald de Jong considers the challenge most problematic for governments. For how can government authorities handle noise nuisance if, as noise experts claim, this annoyance varies "over time rather quickly, under the influence of mood, motivation, [and] situation?" "They can hardly be expected to change standards every five years or so, or to use different standards for different areas" (de Jong 1990: 107–108).

The third case for the persistency of the noise problem—Western culture's fear of silence and its visual character—seems to be the most all encompassing. As the musicologist Bruce MacLeod remarked, "our society seems to be deathly afraid of silence, even though we rarely experience silence in even an approximately pure form." And it is this general fear of silence that explains the omnipresence of background music (MacLeod 1979: 28). The well known example of British BBC accountants who had brand new double-glazed windows, noiseless air-conditioning and silent personal computers installed in their offices is illustrative. Although all the changes were effective in abating noise, the BBC employees felt uncomfortable. They reported feeling lonely and of being afraid that everyone was listening in on their phone calls. This made the BBC decide to buy an expensive noise machine to drive out the silence by producing a continuous and unintelligible hum (Lawaaimachine 1999).

Like MacLeod, the Dutch philosopher Ton Lemaire explains these kinds of responses from a general fear of silence. People are used to surroundings "in which all sounds get produced by humans or by human means and machines" (Lemaire 1995: 107). The right to silence is therefore extremely difficult to protect. Unlike realizing smoke-free spaces, the recognition of the right to quietude would require a basic change in social structure. In our society, driven as it is "by utility, action and the wish to control," listening is considered to be "too passive" (Lemaire 1995: 108). Or, as a journalist puts it on the opinion page of the *Independent:* "Unfortunately, our culture has linked loudness with enjoyment" (Bronzaft 2002: 3).

A variation on the character-of-culture-theme is the supposition that our culture is a visual one, whereas in the past hearing had a higher status. "The ancient Greeks," as Raymond Murray Schafer, initiator of the World Soundscape Project, claims, "were much better listeners than today's architects and acoustical engineers" (Schafer 1994: 13). To his indignation, even the acousticians illustrate their work with slides and charts rather than with sounds. "Yet it is precisely these people who are placed in charge of planning the acoustic changes of the modern world" (Schafer 1994/1977: 128). Our culture, Schafer's colleague Barry Truax recapitulates, tends "to trade its ears for its eyes" (Truax 1978: v).

Schafer finds hope for a better future in Marshall McLuhan's 1962 analysis of a receding print culture: "As our age translates itself back into the oral and auditory modes because of the electronic pressure of simultaneity, we become sharply aware of the uncritical acceptance of visual metaphors and models by many past centuries" (McLuhan, quoted by Schafer 1994/1977: 128). For Schafer, the implications of McLuhan's view are that "as we increase our dependence on acoustic signals, we become out of sheer necessity more conscious of our general sonic environment" (Schafer 1994: 115). Schafer believes that Western culture originally relied on the "inward" drawing ear and views the subsequent rise in the primacy of the "outward" looking eye as a step backward (Schafer 1967: 2). Yet, he continues to believe that our culture may return to giving the ear primacy in the near future, and, if so, the quest for quietude has a far better chance of being successful.

In *The Audible Past,* Jonathan Sterne intelligently comments on this story line, which he calls the "audiovisual litany." In this litany, which predominates in the literature on the senses, hearing is treated as the better sense since it is the "inner" one. While seeing creates distance, focuses on the superficial, and calls on the intellect, hearing surrounds us with sounds, penetrates deep into the heart of the matter, and is inclined to the affective. The religious overtones of this view are obvious, with the eye (the dead letter) in the role of the fallen angel and the ear (the living spirit) as our future paradise. Yet why would the ear be a better sense than the eye? And why should the history of the senses be "a zero-sum game, where the dominance of one sense by necessity leads to the decline of another sense" (Sterne 2003: 16)? The cultural geographer Paul Rodaway views the issue similarly. The new auditory mode is not "a revival of something long since lost, but rather yet another redefinition of the role of the sense of hearing . . . in geographical and social experience" (Rodaway 1994: 114). And as contemporary ethnographers have stressed, today's audio technologies help many people in the auditory control and aestheticization of a meaningful everyday life (DeNora 2000; Bull 2000; Bull and Back 2003).

The idea that our visual culture has atrophied the ear and keeps us from careful listening is still around, however. In some versions, the dominance of seeing and the way it separates subject and object even leads to an aggressiveness that "feeds" noise (Berendt 1985: 17). The key to this line of argument is that we will keep making noise as long as the visual realm continues to prevail in our culture.

Most of these arguments rightly point out various issues that, taken together, make noise problems hard to tackle. Many will only agree that most people in Western societies want economic growth and mobility, that our sense of hearing will always be subjective to some degree, and that silence indeed refers to the absence of life. Many of these arguments, however, also have a decidedly historical and contextual dimension to them, and their validity has changed over time and from place to place. If today, economic growth and increasing mobility, and hence more noise, are seen as belonging to one and the same process, a century ago, the Austrian ethnologist Michael Haberlandt consciously differentiated between the sounds of work and of mobility. He could tolerate "the resonance of work," such as the "song of the hammer, the shriek of the saw, the beat and clatter of the workshop, the stamping of machines," remarking that "we live on that money, don't we?" Yet he lamented the "deafening, enraging noise of the alley," which he described as "a mix of the rattle of carriages, of ringing and whistling, of the barking of dogs and the ding dong of bells with a hundred indescribable overtones that submerge in the uproar" (Haberlandt 1900: 178). Similarly, how hearing is understood to be subjective has changed over time, as have its connotations. Therefore, we should historicize the explanatory force of this subjectivity of hearing. Moreover, the claim that our visual culture is exclusively accountable for noise overlooks the fact that in the early twentieth century, railroad stations were portrayed as having become less noisy *after* the introduction of visual signals that replaced aural ones. Even the argument focusing on the Western fear of silence has its problems. For if we explain noise from a dislike of silence, tautology is just around the corner.

In referring to the ways that noise problems have been defined in the past, such as Haberlandt's lament, I offer a prelude to what will be taken up in subsequent chapters. For the moment, however, I want to show the benefits of using an historical approach for understanding the persistence of public problems about noise by comparing it with the history of stench. This will be the topic of the next section.

NOISE COMPARED TO STENCH

In *Village Bells: Sound and Meaning in the Nineteenth-century French Countryside,* the historian Alain Corbin notes that in the nineteenth century, "hostility to noise . . . was much less discernible than the anxiety aroused by unpleasant odors (1999/1994: 299)." How that came about is the topic of his other famous book, *The Foul and the Fragrant: Odor and the French Social Imagination* (1986/1982).

The years between 1750 and 1880, Corbin argues, were crucial in the deodorization of Western Europe in general, and of France in particular. Intriguingly, those decisive years in the creation of hygiene predated the scientific contributions of Pasteur on bacteria as the cause of disease. From the mid-eighteenth century onward, Corbin claims, the stench of excrement, mud, and cadavers increasingly created panic. This happened first among the social elite, after which it gradually spread to broader segments of the population. Before that time, most people had considered stench as an inevitable aspect of life, a nuisance that was part and parcel of slaughterers', skinners', and tanners' work. Bodily odor was even a sign of vitality and sexual strength. Yet, in the second half of the eighteenth century, medical experts increasingly stressed that stench could do enormous harm to people's health through its miasmas, the infectious substances of the exhalations of body and soil. Smell thus came to be defined as both the symptom *and* the cause of infectious disease and epidemics, the harbinger of death and decay. Also new was a concerted effort by physicians and hygienists to systematize the study of odor. They started to collect airs and gasses, identify their composition and effects, and use their noses to point out the dangers of the offensive smell of rot and decay.

Their work signified a collective refinement in the sensitivity to smell and a growing intolerance to stench. Although they certainly played an important part in the distribution of ideas, hygienists and physicians merely translated the sensitivity of their contemporaries into research. The elite lived in constant metaphysical fear, continuously alert to the processes of dissolution within the body. It was no coincidence that stench came to be associated with the depths of hell. Danger was everywhere, but particularly in the humid holes of the soil, the steam of mud and moors, in the killing stench and fume of city cesspools, and in the repugnant smell of slaughterhouses. Moreover, odor came to be invested with social imagery that made the nose an instrument of social politics.

Subtle changes in policies, Corbin stressed, show how the sense of smell increasingly played a role in the refinement of societal boundaries and practices.

At first, remedies against dangerous smell focused on ways to de-poison the air, and to create flavors and perfumes that restored the balance of good and bad airs. Experts proclaimed that strong, bestial odors such as musk and amber were best suited for the abatement of persistent stench. As soon as the idea that filth obstructed the pores had been accepted and the first steps of maintaining bodily hygiene had been taken, however, the attitude of the elite toward musk and amber changed. The use of strong perfumes increasingly incurred the suspicion of a lack of hygiene. The growing sensitivity to bodily odors made both lighter, flowerlike perfumes and the cultivation of one's sensitivity to smell more popular. The social elite were also increasingly astonished by the tolerance of the common man for stench. The more sensitive one's nose, the more refined one's nature—a refinement, the elite noted, that workers did not possess.

Similarly, the attitude toward the masses shifted. Initially, a general distrust of the emanations of crowds existed, particularly of those packed up in closed spaces, such as in hospitals, prisons, barracks, ships, churches, and theaters. Since humidity was considered dangerous and movement purifying, numerous ventilation systems for crowded spaces were invented, which was further stimulated by a refinement of the analysis of air that showed a decrease of oxygen in closed spaces with lots of people. In the course of the nineteenth century, however, not a dislike of the fumes of undifferentiated masses, but a dislike of the smell of the poor began to predominate. Pauperism and stench were seen as one and the same issue. The elite distinguished itself from the rest of the population on the basis of smell, which underlined the danger of contagion and justified the need to discipline and subject the lower classes. Therefore, after the citizenry had deodorized its own bodies and houses, the semiprivatization of toilets (located on the porches of tenements) and the inspection of the public's houses to eradicate smell grew increasingly important. Thus the focus of attention shifted from the biological to the social and from public to private space.

In addition to ventilation, paving, drainage, and creating space between people, there were significant strategies for reducing smell. Rules were introduced for the clearing and cleaning of cesspools, streets were repaved and broadened, and waterworks, latrines, and the regular changing of clothing in hospitals became common. Within private space, filth had to be covered up with lime,

15

chimneys became mandatory, and bathing gradually developed into a routine. The introduction of substances such as chlorinated water and zinc chloride and swan-necked containers was of enormous help. Within a context of increasing utilitarianism, the possibilities of using human excrement as fertilizer and animal remains for salt ammoniac were increasingly recognized, albeit with different effects. For example, in France the use of human excrement as fertilizer initially held up the planning of sewage systems, whereas the collection of carcasses contributed to the deodorizing of public space.

In the early nineteenth century, the idea that smell caused disease influenced even the nuisance regulation of industries. Damp emanations produced by the rot of collected animal and vegetable materials were considered to be unhealthy and inconvenient in terms of the law. Yet chemical effluences were not seen as dangerous. Therefore, the process of industrial deodorizing proceeded slowly (Corbin 1995/1991: 156).

We can infer from this history of stench that, first, the identification of smell as a cause of disease, and thus a public health threat, contributed significantly to the kinds of interventions chosen and solutions found for stemming stench. Hygienists played a significant role in this process. Second, it is important to note that the stress on differences in sensitivity to smell contributed to, instead of hampered, the abatement of smell. This occurred because smell became the sign of social difference and hierarchies that, together with the idea that smell caused contagion, legitimized interventions, such as home checks and instructions on hygiene, in the lives of lower class people.

As Andrew Aisenberg has shown, the legitimization of the intervention in private spaces in Paris was based on Pasteur's scientific theories from the late nineteenth century onward. The prevention of disease became increasingly associated with the disinfection of the home, the isolation of diseased patients, and tracing the persons with whom they had had contact in order to find a contagion's source. Since respect for the integrity and autonomy of the family would endanger society, the danger embodied in the contagion's source, "made the acceptance of social duties, articulated and presented by a regulatory authority, an integral part of what it means to be an individual in urban space" (Aisenberg

---

1999: 173). If in the late nineteenth century, then, the conflict between liberty and social order was legally resolved by the scientific notion of contagion, at the century's start the mere reference to the miasmatic smell of the poor as a threat to social order had been sufficient grounds for intervening.

Third, it is remarkable that considerations of economic gain hampered as well as stimulated the abatement of smell. The use of excrement as fertilizer hindered the installment of sewage systems in France. But as soon as animal remains were understood to be a source for the production of new and useful substances, people stopped allowing them to rot and stink.

Many of Corbin's observations about the social meaning of smell are equally applicable to noise, as this book will amply demonstrate. Just as with stench, noise was also considered to threaten the social order. If stench became symbolically associated with the depths of hell, noise became characterized as infernal din. Social elites not only considered the lower classes to be insensitive to smell and bestial odors, but also portrayed them as being indifferent to noise (Schafer 1994/1977: 223). However, the public problem of noise has been resolved less thoroughly than the public problem of smell. This calls for a rephrasing of my questions. For one thing, why have we not become so afraid of or touchy about sound as we have become about smell, the putative health menace number one between 1750 and 1880? Why did the social hierarchy of sound prove less helpful in the abatement of noise than the social hierarchy of odor had been in the case of stench? And which developments undermined the association of silence and economic gain strongly enough to be a mainstay in the fight against noise?

WAYS OF UNDERSTANDING THE HISTORY OF NOISE

Having used the comparison between stench and noise to gain a better understanding of the significance of the *definition* of such public problems—their putative causes, consequences, solutions—I will return to the work of Joseph Gusfield. His work is one of my sources of inspiration for how to examine the history of public problems of noise. Drawing on the introduction of social

constructionism in the theory of social problems in the early 1970s, Joseph
Gusfield developed a framework for studying the culture and structure of public
problems (Gusfield 1981: 4; Miller and Holstein 1993). The distinction between
their cognitive and moral dimensions was crucial to Gusfield's understanding of
public problems. "The cognitive side consists in beliefs about the facticity of the
situation and events comprising the problem—our theories and empirical beliefs
about poverty, mental disorder, alcoholism, and so forth. The moral side is that
which enables the situation to be viewed as painful, ignoble, immoral. It is what
makes alteration or eradication desirable or continuation valuable. . . . Without
both a cognitive belief in alterability and a moral judgement of its character, a
phenomenon is not at issue, not a problem" (Gusfield 1981: 9–10).

Equally important are the notions of *ownership, political responsibility,* and
causal responsibility. Owners are those groups or institutions defining the problem,
whereas those in charge of actually solving the problem by intervening are
the ones with *political responsibility.* In contrast, *causal responsibility* is about the
explanation of phenomena, such as saying that the source of impure air is the
automobile. Yet the interesting thing is that the structure of public problems is
often "an arena of conflict in which a set of groups and institutions . . . compete
and struggle over ownership and disownership, the acceptance of causal theories,
and the fixation of responsibility" (Gusfield 1981: 15).

Gusfield's motivation for writing his book was to show how the many
professionals involved in trying to solve the "drinking-driving problem" were
locked into a specific definition of that problem. This definition comprised the
following: "Alcohol leads to impaired driving and increases the risk of accident,
injury, and death. Since drinking coupled with driving 'causes' auto accidents,
solutions lie in strategies which diminish either drinking or driving after drinking" (Gusfield 1981: 7). The key to this problem definition was the idea that
alcohol and car safety are first and foremost the problem of individual motorists,
more specifically, the problem of "the conflict between self-control and self-
indulgence" (Gusfield 1981: 173). "Two things struck me as especially significant
by their absence: the lack of involvement of alcohol beverage distributors—bar-
tenders, sellers, manufacturers—and the inability or unwillingness of people to

see the problem of drinking-driving as a problem of transportation" (Gusfield 1981: 7). Several alternative definitions had been proposed in the course of time. One of these defined safety as a problem related to the construction of the car, for which the car industry had responsibility. Yet such a conception failed to become predominant in the episode that Gusfield brought to the fore, which he relates to the high status of individualism within American culture.

Gusfield's focus as a sociologist was on the culture and structure of public problems per se. The historical dimension of his research had to underpin his method of irony by which he meant "to hold up that which is taken for granted, familiar, and commonplace as something strange and problematic" (Gusfield 1981: 191). "To find alternative ways of seeing phenomena," he added, "is to imagine that things can be otherwise" (Gusfield 1981: 193). And since he could partly find such alternative ways of seeing in the past, historical research was an important entry into the study of public problems. In this book, however, the historical dimension has an additional use. Following the sequence of public problems of noise and their changing structures over time allows me to show how strategies for the solutions of noise—such as individualizing, objectifying, and materializing noise—were transferred from the definition of one type of noise problem to another. At times this created situations in which the people abating noise tried to characterize and win the ensuing war with the former war's weapons. Similarly vital to this book is the branch of public problems theory that studies the moment of *discourse coalitions* in the emergence of public problems (Hajer 1995). How did social movements that defined noise problems involve other social groups in their strife for tranquility?

Studying the contribution of law and science to the *staging* and *drama* of public problems is also highly significant for answering this book's key questions. Gusfield analyzed juridical documents and public presentations, even scientific ones, "as performances—as materials which dramatize the drinking-driving phenomenon as both a cognitive and a moral matter" (Gusfield 1981: 18). State laws, for instance, "hold the individual and not the auto industry or the road or the locality 'responsible' for accidents." Such laws are based on deterrence, the idea that "the individual motorist can be led to more diligence in driving through the fear of police apprehension and legal punishment" (Gusfield 1981: 45). The ways in which scientific information on drinking and driving was gathered, and how these facts were presented and classified have been crucial to Gusfield's project as well. Similarly, the delineation of responsibility for acoustic privacy in law is paramount to the understanding of the history of noise. And so are the processes leading to the rise of the decibel in the measurement of noise, and the significance of these measurements to the definition and dramatization of noise.

Thus, this book aims to follow the changing order of public problems of noise over time and give a special ear for the contribution of science and law to the drama and definitions of these problems. This means that in addition to public problems theory, the field of Science and Technology Studies (STS) is indispensable to this study. STS contributes to my analysis in many distinct ways, but studies of *standardization* will appear in this book more than once. For example, what did the prevailing definitions of noise problems mean with respect to the character, acceptability, and employability of the standardized units and technological tools of noise measurement? What did these units entail in the interventions in noise problems? And what does the investment in standards as techniques of trust, coordination, and control say about the role of *experts* in the technological culture of the twentieth century?

What makes this study different from much of the current work in STS is not merely the connection it creates between STS and public problems theory, but also its aim to acknowledge the general public's acceptance of *technological determinism,* the idea that technology develops autonomously and simply takes society by surprise (Smith and Marx 1994). To do justice to their acceptance of technological determinism, this book will analyze the effects of situations in which, at least to the general public, new sounds seemed to fall out of the sky—in a rather literal sense in the case of aircraft. In doing so, this study can fully harvest the fruits from the social and radical constructivist studies that stress the co-evolution of science, technology, and culture (Bijker, Hughes, and Pinch 1987; Latour 1987), while taking the consequences of belief in technological determinism seriously.

The third field of considerable input is the history, anthropology, cultural geography, and philosophy of the senses. The better studies from these disciplines help me to reckon with the historicity "of the modalities of attention, thresholds of perception, significance of noises, and configuration of the tolerable and the intolerable." Again, this quotation comes from the work of Corbin. Corbin contrasted his approach to that of Guy Thuillier who attempted to "to compile a catalogue and measure the relative intensity of the noises which might reach the ear of a villager in the Nivernais in the middle of the nineteenth century" (1995: 183). Indeed, while Thuillier considered the history of noises to be part of the history of mentalities, what he ended up doing was summing up the variety of sounds that were audible to villagers. These included the laments of death's harbinger, as well as the sounds of forges, steam machines, and telephones (Thuillier 1977: 231–234). Corbin certainly valued this type of work. "It aids immersion in the village of the past; it encourages the adoption of a comprehensive viewpoint; it helps to reduce the risk of anachronism." Yet it wrongly implied the idea that "the habitus of the Nivernais villager of the nineteenth century did not condition his hearing, and so his listening" (Corbin 1995/1991: 183).

Such a critique also partially applies to the work of Raymond Murray Schafer, the Canadian composer, environmental spokesman, and author of *The Soundscape: Our Sonic Environment and the Tuning of the World*—the book that resulted from the World Soundscape Project. This project started in the late 1960s and involved both education and research. Schafer's first interest was in noise pollution. He soon relabeled this as soundscape design, however, because it had a more positive tone and was more popular among students than noise abatement (Schafer 1969; Schafer 1999; Järviluoma 1994). In Schafer's view, several routes could lead to a more pleasurable *soundscape,* or sonic environment. The first was mapping historical and recent changes in the world's soundscape. The second was promoting "ear cleaning" workshops and "soundwalks" to enhance the public's listening capabilities (Schafer 1967). The third was employing environmental sound recordings as samples for new music compositions. In the 1990s, Schafer started presenting his ideas for acoustic design in public spaces. One proposal was to organize the aural information at railway stations and airports in such a

Figure 1.1 Logo of the World Forum of Acoustic Ecology. This logo was derived from an art work by Liliane Karnouk, originally created in 1978 for a poster announcing the radio program *Soundwalking,* produced by sound artist Hildegard Westerkamp for Vancouver Co-operative Radio. Courtesy Liliane Karnouk.

way that different kinds of information had distinct sonic motifs, an approach that is analogous to what occurs in nature, where, for example, each species of bird has it own sound (Schafer 1993: 24). The proposal expressed Schafer's eco- logical approach of sound. A similar passion for the acoustic richness of public spaces and consumer products, rather than for mere quantitative noise control, can be found in the World Forum for Acoustic Ecology,[6] an offshoot of the World Soundscape Project. (Figure 1.1.)

The larger part of Schafer and his colleagues' work, however, was con- cerned with documenting changes in soundscapes by archiving and describ- ing sounds through sound recording journeys and sound diaries (Schafer 1977a, 1977b; Truax 1996). Schafer designed parameters like the "soundmark," a unique "community sound," which included sounds such as the slam-and-slick of doors in the Paris Metro and "keynote sounds." Keynote sounds were sounds "heard by a particular society continuously or frequently enough to form a background against which other sounds are perceived," such as the sound of geysers on Iceland or the murmur of electric equipment in modern dwellings (Schafer 1994/ 1977: 274, 272). In addition, Schafer developed an extensive sound notation system, including the physical characteristics of sound (such as duration, frequency, and dynamics) as well as referential aspects (meaning, purpose, and function). All of this helped him to document a shift from a preindustrial "hi-fi" sonic environment, in which signals are clearly audible, to an industrial "lo-fi" soundscape, in which individual sounds are masked and overcrowded. One of Schafer's colleagues, Barry Truax, developed an even more distinctly ecological perspective by pleading for the preservation of sonic "variety" (Truax 1984: 97).

Although Schafer definitely sought to understand the rise of the lo-fi soundscape, which he attributed to the power of industrialists, his project and its aftermath remained descriptive rather than explanatory.[7] His conclusions even suffer from the same kind of fallacy that Corbin identified in Thuillier's work. To Schafer, the character of the sonic environment is what changes, not the ways in which people listen to it. Such an approach, however, makes it hard to understand why new mechanical sounds, as I will show in later chapters, often seemed to be perceived "unmasked," as if these sounds arose in a hi-fi surrounding instead of in the lo-fi sonic environment that Schafer assumes exists in industrial society. I prefer Emily Thompson's definition of the soundscape, as published in her widely and rightly praised study, *The Soundscape of Modernity*. In Thompson's view, the soundscape is "simultaneously a physical environment and a way of perceiving that environment; it is both a world and a culture constructed to make sense of that world." Her physical environment, moreover, does not only encompass sounds, "but also the material objects that create, and sometimes destroy, those sounds" (Thompson 2002: 1). What's more, Thompson's approach reduces the downsides of the "visual analogies" of Schafer's notion of soundscapes, a shortcoming for which both Paul Rodaway and Joy Parr have criticized him (Parr 2001: 736). Such visual analogies may lead one to construe a static image and to forget about the "dynamic" aspects of the auditory experience (Rodaway 1994: 86–87).

This is not to say that the work of Schafer and other soundscape theorists is not informative or inspiring. On the contrary, their documentation of soundscapes is phenomenologically rich and their critical program evocative. Yet this book is a study of shifting public problems of noise. And to study these changes, I need to be thoroughly historical. This means that the physical characteristics of sound are not sufficient to understand why particular sounds came to be defined as noises or why private problems of noise became public ones. These questions require acknowledging transformations in the ways people listen to sounds and their cultural meanings. Being thoroughly historical also requires putting the three most common explanations for the persistence of noise—economic growth, the subjectivity of hearing, and the sensory priorities of our culture—in parentheses. The historiography of stench shows us that economic gain and solving sensory problems can go together in certain contexts and that the notion of sensory subjectivity may help instead of hamper in tackling such problems. Moreover, critical studies of the senses stress that a visual culture is not by implication less attentive to sound. Yet unlike stench, noise never acquired a robust enough status of societal danger to make deeply encroaching and lasting interventions appear to be legitimate. This provides the first clue to understanding why today's columnists only dare to discuss noise with a touch of irony. What has contributed to this situation? And what needs to be done to understand the persistence of public problems of noise?

Listening to Technology through the Ears of the Past

Since this book's focus is on the decades immediately following the efforts to raise issues of noise actuated by new kinds of mechanical sound—industrial, traffic, audio, and aircraft sound—it does not fully cover a century of noise abatement policies. Instead, it aims to scrutinize the most fervent episodes of public debate and action so as to understand what's behind the persistence of noise as a public problem. These episodes succeeded each other roughly between 1875 and 1975. By 1875 most European countries had nuisance laws that would influence the definition of future noise problems. And by 1975 most European countries had designed or introduced encompassing laws on noise nuisance or environmental legislation in which noise was included.

Nor does the book cover all of Western culture in each chapter. The scope of primary sources concerning music, medicine, and acoustics spans North America and Western Europe. Throughout most of the twentieth century, these worlds, notably the journals and meetings that constitute their inner circles, were highly international. With respect to other relevant sources, such as pamphlets, essays, and newspaper articles on noise, the archives and journals of national noise abatement societies and the archives of national research institutes, governmental committees, and national policy documents, my main concern has been with Western Europe, notably the United Kingdom and Germany, as the first two European countries with noise abatement societies, and the Netherlands. In addition, I consulted this subject's most significant primary sources from the United States. In the mid-1970s, the Dutch government considered the United States, the United Kingdom, and Germany as the countries "furthest ahead" in the legislation on noise abatement.[8] Yet even in these countries, the public debate about noise all but died down. This contrast makes these countries particularly relevant for studying the persistence of public problems of noise.

Because the various national noise abatement societies were very eager to report about antinoise activities elsewhere, while governmental committees usually turned to other countries for information before deciding what to do, their documents provide a wealth of information about the other Western countries. Still, subsequent chapters will foreground some countries more than others, and this is in part motivated by the fever of discussion, by how developments in one country triggered changes elsewhere, or by the extent to which a particular type of noise, as in the case of neighborly noise, requires extensive local detail.

At first sight, one might expect this study to be accompanied by a compact disc with historical recordings or contemporary recordings of historical artifacts. This, it seems to me, would not be very helpful. Listening to a recording of museum steam machines might give you the impression that these machines were not very "loud" at all, forgetting that steam machines may not have been as well oiled when originally in use as when in use in a museum decades later.

Hearing the cracks and noises of a phonograph recording may initially enlighten their historical status as "mechanical" instruments. Yet, the very same sounds complicate our understanding of the "tone tests" of the early twentieth century in which audiences were unable to hear the difference between performers and records playing (Thompson 1995; Siefert 1995; and Sterne 2003)—something that is hard to believe today. When presenting papers on the history of traffic noise, I have often confronted my audience with a recording of the actual sound of an automobile horn from the 1930s. Even if my act always managed to make the audience laugh because the recording sounded like a dog with a cough, it failed to contribute to its grasp of why honking was such a dominant public problem in 1930s. Thus, if we take seriously the historicity of perception, as well as of public problems, recordings are a far less informative or a much more complex source than one might think (Smith 2003a, 2004).

Our challenge, then, is to historicize the sensory experience of sound, and to listen to the sounds of technology through the ears of those people who complained about these sounds. Subsequently, we have to understand how the complainants were able to dramatize these sounds in such a way that they could be packaged into noise problems to be discussed in the public arena. What repertoires for the dramatization of sound could the complainants draw from?

INFERNAL DIN, HEAVENLY TUNES:
REPERTOIRES OF DRAMATIZING SOUND

THE ARCADIAN MILL

In the first decade of its existence, the British Anti-Noise League often chose pastoral scenes such as a sailboat on a meandering river, a serene lake surrounded by trees, a tranquil sea, a winter landscape, or a bridge over a small brook for the cover of its magazine, *Quiet*.[1] One of the scenes featured a windmill with a shepherd and his sheep in the foreground. No barking dog was to be seen. (Figure 2.1.) The message was clear: *Quiet's* covers depicted the tranquil life that the British Anti-Noise League longed for. It believed that nature was without noise and that pastoral life was paradise.

The British Anti-Noise League chose these illustrations for their magazine's covers because they alluded to the antinoise activists' love of silence, to a serene and still life. The Arcadian mill was the perfect icon of a restful Eden on earth. Today the windmill remains a symbol of a peaceful life. However, its symbolism bears no relationship to how the sounds of wind and water mills have been valued in the past. In early modern towns, these mills were common objects of noise abatement (Poulussen 1987). The fact that we now consider mills a symbol of serenity is a consequence of viewing preindustrial society as quiet *compared to* industrial and postindustrial societies. Several academics

Figure 2.1   Front page of the magazine *Quiet* 1, no. 3 (Autumn 1936). *Quiet* was issued by the British Anti-Noise League.

have assumed that there has been a gradual crescendo in the history of Western sound over time, evidenced by an increase of the loudness of everyday sounds and a deteriorating signal-to-noise ratio. Literary scholar Bruce Smith recently claimed that the acoustic world of early modern England never reached above sixty decibels "apart from barking dogs and the occasional gunshot," whereas the sound level produced by average city traffic today is eighty decibel and that by jets 140 decibel (Smith 1999: 76, 50).[2] Acoustic ecologists have underlined that all our present-day sounds "are running together," making it difficult to orient ourselves in time and space. In contrast, the environment of the preceding ages had enabled people to "hear things clearly" and identify "the relationships of the various sounds."[3]

Such claims about our sonic past and present invite us to associate the mill with tranquility. Its identification as a symbol of serenity however also depends on its setting in scenes such as on *Quiet*'s cover. The mill is set in a pastoral rather than an urban scene, and the spectator is placed at a distance from the mill. The mill's solitude reduces, on paper at least, the potential disturbance of the sounds it makes. The mill's double stylizing—its setting in pastoral life and its depiction as a solitary source of sound at a distance—is a sign of the importance of drama in packaging the problem of noise.

In his study of the drinking-driving problem in the United States, Gusfield claimed that we "live in a forest of symbols on the edge of a jungle of facts" (Gusfield 1981: 51). What he meant is that individuals cannot experience the aggregate level of public problems. Instead they depend on experts of several different kinds—civil servants, academics, journalists, pressure group lobbyists— to organize and dramatize public problems in order to make these problems comprehensible to them. In Gusfield's analysis, the symbolism invoked and the cognitive structure of the problems sustain each other. Fiction and drama are part and parcel of the ways in which knowledge is presented.

Thus, groups that are looking for an audience for a public problem need to create a sense of urgency for the issue. This can be done in many ways, for instance by circulating fact-sheets describing alarming numbers of victims, but also through other rhetorical forms, such as the ideal of the tranquil life that

British Anti-Noise League conveyed through their image of the Arcadian mill. The people who aimed to put noise on the public agenda had to translate everyday complaints about noise into a serious drama in order to make their charge convincing.

In this chapter I want to create an eye for the *repertoires of dramatizing sound* such people had at hand to use. Which conventions of persuasive speaking about sound were culturally available to those staging noise problems? I introduce these conventions in two ways. First, I review recent literature on the socio-cultural history and anthropology of sound to explain the most predominant cultural connotations of noise and silence prior to the early twentieth century, the period just preceding the establishment of the first noise abatement societies. This overview will show that the right to define sound as forbidden noise, the act of making loud sounds, and the plight of maintaining silence all refer to deeply embedded cultural patterns and hierarchies during this period. Second, I draw up an inventory of *auditory topoi of technology,* a set of stereotypical expressions of mechanical sound, drawn from a sample of quotations from Western novels published from the period prior to or contemporaneous with the rise of the noise abatement societies.

Both the literature review and inventory of expressions of mechanical sound enable me to identify in subsequent chapters which connotations and conventional phrasings of sound were employed to dramatize sound. Moreover, the overview of repertoires provides the instruments necessary to understand the reciprocal reinforcement of the dramatization and the definition of noise problems. Finally, it helps me to analyze the effect of particular forms of dramatization in the specific contexts in which they were put forward. As subsequent chapters will make clear, the kinds of dramatization that complainants and anti-noise activists chose did not always match the types of dramatization required in the contexts in which they attempted to convince their audience.

One could easily misread this chapter's purpose, however, as taking dramatization of sound as part of a timeless substructure of culture that determines the course of events in debates about noise. This, however, is not the aim of this chapter. The review and inventory are intended to function heuristically, so that repertoires of dramatizing sound and their effects can be identified in subsequent chapters. By this I do not wish to imply that noise and silence have immutable connotations or that the relationships between literature and public problems are one-way and clear-cut. Those who defined public problems concerning noise often echoed older connotations of noise and silence, but they also gave these meanings new twists. Those who complained about noise may or may not have read the Western novels I analyze. In turn, these novels may or may not have been informed by the ideas of noise abatement campaigners. These issues, however, are not the ones being discussed here. This chapter is about getting a feel for the ways in which sound was dramatized—ways that were, I argue, "around" culturally before and at the time antinoise activism entered the scene.

CULTURAL CONNOTATIONS OF NOISE AND SILENCE

"Breaking the sound barrier" is the well-chosen title of a 1996 article in which the historian Peter Bailey aims to bang "the drum for noise." A general history of noise, Bailey claims, should start with mapping out changes in the production of noise in terms of its "means, type and volume" (Bailey 1996: 55; Bailey 1998: 201). Yet, Bailey adds, it should also focus on the subjective and culturally determined perception and response to noise, as well as on the changing hierarchy of sensory perception. Showing important temporal, class, gender, and age differences in relation to noise attitudes and stressing the increasing dominance of visual over auditive perception, Bailey largely lives up to the task with which he charges historians of noise. Most relevant are his notes on the conventional connotations of noise and silence in Western culture. This section focuses on the history and anthropology of such connotations, or, put more precisely, on the symbolism of noise and silence and its relation to technology, by discussing Bailey's as well as others' contributions to these two fields of research.

Before doing so, however, it is necessary to explain my use of the phrase *symbolism of noise and silence*. In linguistics, the term "sound symbolism" is used to denote sound units, such as phonemes and syllables, when they express meaning beyond their linguistic function. The relationship between sound and meaning can be iconic, indexical, or symbolic. The relationship is iconic when a sound unit's meaning seems naturally linked to its articulatory or acoustic characteristics. A fascinating yet disputed example is the claim that high frequency sounds such as /i/, which have comparatively short wavelengths, are often associated with ideas of smallness, such as in the word "baby." A sound bears an indexical relationship with its meaning when the way in which it is bodily produced makes sense, for instance, when a higher pitch, related to greater vocal cord tension, is indicative of a heightened emotional state. The relationship is considered symbolic when the link between a sound and an idea is conventional and has no apparent motivation (Nuckolls 1999: 228–229). It is my interest in such conventional links that has inspired the literature review that follows. I will not focus on sound symbolism in language proper, however, but on sound symbolism in culture, and especially the symbolism of noise and silence.

Let me start with noise. Bailey distinguishes three types of socially defined noise: Noise as merriment, such as laughter; noise as embarrassment, such as a fart; and noise as terror. The last type of noise has been discussed most frequently in the literature. As Bailey and other authors have made clear, noise—varying from shouting to shooting and bombing—has always been an important ingredient of warfare. This kind of noise can be used to encourage persons to steel themselves against an oncoming threat as well as to terrify the enemy. The quotation below, taken from Alessandro Scarlatti's late-seventeenth-century oratorio *Giuditta* [Judith] illustrates the use of noise as a form of terror by describing what happens if one wages war *without* producing loud sounds. Judith wants to withstand the marching enemy, and bewails the silence of the drums within her camp:

Trombe guerriere, perchè tacete?
S'ogn'alma ingombra
di tema un'ombra,
voi col tacere
piu l'accrescete.

[War drums, why are you still?
If a shade of fear
cramps any soul,
you add to this
by remaining silent.][4]

The historian Richard Cullen Rath, on the other hand, has elucidated the terrifying effect of war's noise on the enemy. He describes the shouts and cries
that the First Nations tribesmen used as a "standard way of beginning a battle"
in seventeenth-century North America. For contemporary colonists, these yells
sounded like "horrible shouts and screeches, as though so many infernal hell-
hounds could not have made them more horrible" (Rath 2003: 153).

The use of noise as a form of terror has also been manifested in the
"symbolic violence" of crowds, which is a sort of "counter-terror" employed
to "intimidate, shame or ridicule enemies of the people and offenders against
traditional values." Rough music, "*katzenmusik,*" or kettlemusic, a cacophony of
noises produced by shouting, laughing, chanting, and the drumming of pots and
pans could, for example, be used to embarrass newlyweds (Bailey 1996: 52–53).
In less innocuous versions, kettlemusic was used to frighten persons accused
of socially unacceptable behavior (Kruithof 1992: 48; Smith 1999: 154–155).
Whatever the context, Bailey argues, rough music has often been taken as "the
sound of disorder" (Bailey 1996: 53).

For this reason, drums, horns, and bagpipes have occasionally raised suspicion (Libin 2000). Rath has documented how American slaveholders banned
the use of high-volume sounds such as the drums used by enslaved African
Americans in the seventeenth and eighteenth centuries. As a consequence,
slaves shifted to quieter instruments, such as fiddles "that helped them build
and maintain autonomous cultural spaces under duress" (Rath 2003: 173). And
Mark Smith has shown that an 1845 Georgia statute prohibited slaves from using
drums, horns, or other loud instruments with which they might communicate
their "wicked designs and intentions" (quoted by Smith 2001: 56).

The perception that noise could be socially disrupting was widespread. The ban on the slaves' instruments is just one of the many manifestations of sounds having been defined as noise because they signified a threat to a particular social order. The language of foreigners, the drunken turbulence of the poor, the street music of immigrants, and the emotional expressions of evangelical worshippers have all been, at different times and places, characterized as din by the ruling classes and religions (Smith 1999; Smith 2001; Picker 2003; Schmidt 2004). Even the non-native sparrow, introduced in the United States in the mid-nineteenth century, was accused of being an invader that produced "disagreeable chatter" and an "indescribable jangle of harsh sounds" (cited in Coates 2005: 649). If socially disruptive noises could not simply be banned, creating distance or territorializing tranquility was often proposed as an alternative solution. Schivelbusch provides the intriguing example of nineteenth-century continental-European travelers who preferred the compartment train over the American public carriage. Whereas the public carriage confronted them with the sounds and sights of others, the compartment train, an imitation of the coach, would allow them to be left alone. But after two murders were committed on the train that went unheard by passengers, train designers created peepholes, one-sided corridors, and sliding doors to enable passengers to see, but not hear their fellow-travelers, whose sounds they felt were an intrusion of their beloved privacy (Schivelbusch 1979: 89).

Making noise has not only been associated with social disruption, however. The flip side of this association is the power of those who are allowed to make their presence heard. According to Raymond Murray Schafer, the association of noise and power has always been part of the human imagination. Throughout history, societies have known certain types of noise, which Schafer calls *Sacred Noise,* that were "not only absent from the lists of proscripted sounds which societies from time to time drew up," but were in fact "quite deliberately invoked as a break from the tedium of tranquility." Creating this kind of noise—in religious festivals celebrating the harvest, in rituals exorcising evil spirits, in ringing church bells, or in playing the organ—was aimed at making the deity listen. Those in society who produced Sacred Noise, Schafer stresses, not only made "the biggest noise," but actually had "the authority to make it without censure." Where noise was granted "immunity from human intervention," power was near (Schafer 1994/1977: 76, 51). Gods who produced thunder and lightning and priests who used drums and bells are traditional examples of this phenomenon.

Indeed, as Rath illustrates, the inhabitants of seventeenth-century America attributed a "tangible power to the sound of thunder," which the Puritans viewed as the voice of God and Native Americans as the act of divine "thunderers" (2003: 31). Similarly, studies of African cultures show that drum sounds and words are often considered to be laden with a force of their own or the powers of ancestors (Stoller 1989: 110, 121). For the Navajo, to "speak and to sing in ritual contexts is to control air and, at the same time, create order, harmony, and beauty," thus making "contact with the ultimate source of life" (Witherspoon quoted by Nuckolls 1999: 227). According to William H. McNeill, moving rhythmically together in community dance, song, and drill possesses a force of its own through its ability to arouse "warm emotions of collective solidarity" (McNeill 1995: 152; Pacey 1999: 20). Mass gatherings can give people a sense of irresistible strength. This was the case in the past and can still be found today in phenomena such as the yells and songs of sport fans.

Interestingly, Schafer extends his ideas about the links between permissible noise and the possession of power to technology. Sacred Noise, he claims, was eventually transmitted to machines. Its power descended from "God, to the priest, to the industrialist, and more recently to the broadcaster and the aviator." Only this transfer, he believes, can provide a compelling reason for the astonishing fact that the toxicity of industrial noise was not recognized until the late phases of the Industrial Revolution. Like church officials and other authorities in previous centuries, industrialists had been "granted dispensation" to make noise (Schafer 1994/1977: 76). As medical historian Allard Dembe claims, a general "cultural association of loud industrial noise with admired societal traits," such as strength, progress, prosperity, and prowess, hampered the recognition of noise-induced hearing loss as a disorder deserving compensation (Dembe 1996: 203).

Dembe's remarks suggest that the association of noise-as-loud-sound with power-as-strength-and-significance is an important attraction of some technologies. In his study of the meaning of village bells in the nineteenth-century French countryside, Alain Corbin carefully documents the perceived correlation between the loudness of bells and the significance of parishes and municipalities. Bells, he claims, served to "give expression to hierarchies" and also to "defend a territory," "impart a rhythm to time," "identify personages deemed worthy of honor," "make announcements, exhort people to assemble, sound the alarm, and express general rejoicing" (Corbin 1999/1994: 159). It was, Corbin stresses, "as if the powerful ringing of the bell represented a victory over chaos and, for a community, a symbol of cohesion regained; it was an instrument whose sound enabled people to assemble, and it was the sign of a social order founded on the harmony of collective rhythms" (Corbin 1999/1994: 290).

Another sign of social order was the buzz of a town itself, at least to early nineteenth-century Americans "emerging from days in the woods," as they migrated across the continent. To them, these sounds represented a sign of relief, "a heard archipelago in a quiet, howling wilderness" (Smith 2001: 108). And as "Americans pushed westward, the rhythmic, ordered sound of progress quashed the sporadic, unpredictable noise of the frontier (Smith 2003b: 139)." For many during this period, the cadence of rising industry had a similarly positive meaning. This hum stood for activity and wealth. And although elites "did not always revel in registers of progress, ... they nevertheless recognized them as necessary and manageable"—through municipal statutes. In principle, the bourgeoisie warmly welcomed the "clanking of machinery," "the far-resounding echoes of ponderous hammers," and the ringing of factory bells as the sounds of industriousness (Smith 2001: 128).

In many ways the loudness of technology continued to be associated with strength in early twentieth-century Western culture. For instance, in 1923 promoters of the electric car sadly noted that "the rattle and clatter of the gasoline vehicle" impress the man on the street more than "the quiet reserve and staying qualities" of the electric car (Cushing cited in Mom 1997: 475). Men were held to love the din of the internal combustion engine for its expression of speed, risk, and power. In Motor Boys and Motor Girls novels, serial children's books published between 1910 and 1917, Motor Boys preferred "their cars noisy, like racing cars, and even disconnect[ed] mufflers," whereas the Motor Girls liked a different sound. According to one, the "car went noise-lessly—the perfection of its motion was akin to the very music of silence" (McShane 1994: 169).

Such observations and interpretations can best be structured with help of Anthony Jackson's analysis in "Sound and Ritual," a review of the notion of sound in the anthropological literature of Claude Lévi-Strauss, Rodney Need-ham, and others.[5] According to Jackson, man-made noises, or unpatterned, disordered, arrhythmical sounds in ritual "reflect uncontrolled situations or tran-sitional states or threats to patterned social order," while "a taking up of a rhyth-mical beat again reasserts human control over events" (Jackson 1968: 295–296). Although Jackson modestly characterizes his conclusions as "speculation," his findings largely fit the historical and anthropological work mentioned above. These studies have shown that noise as "unwanted sound," whether regular or unpatterned, has often been associated with a disruption of a particular social order, terrifying at times, whereas rhythmic and/or loud, positively evaluated sounds have been associated with strength, power, significance, masculinity, progress, prosperity, and, last but not least, control.

Extending this line of thought, noise as "unwanted sound" has much in common with Mary Douglas's definition of dirt as "matter out of place" in her work on purity. In primitive societies, averting dirt and striving for purity with respect to food, sex, and the body are ways of creating order out of chaos. Moreover, the fight against dirt can be a symbol of as well as a lightning rod for deeper societal conflicts (Douglas 1976/1966). Many academics have applied Douglas's remarks on primitive societies to modern societies as well, for instance by relating a cultural obsession with purity to episodes of rapid modernization (Labrie 1994).[6] Whether obsessions with noise have been more dominant in stages of societal transition cannot yet be determined. Actually, this may be an issue that is almost impossible to settle on a general level. Since which transitions can be considered deep or fast enough to cause a societal obsession with noise? And how should we define a societal obsession with noise: as an expression of public debate and action or as an expression of complaint?

Notwithstanding this unresolved issue, the review above makes clear that noise as unwanted sound, in the centuries before the first peak in noise abatement campaigns as well as in non-Western societies, has often been a symbol of disorder, while welcomed loud and rhythmical sounds has had connotations with diverse varieties of strength. Yet such a symbolism of noise is incomplete without a knowledge of past connotations of silence.[7] When focusing on noise, for instance, one could easily forget that in many religions, contact with gods not only proceeds through rites of making noise, but also through rites of keeping silence (Menninghaus 1996).

In *The Art of Conversation,* the historian Peter Burke devotes a chapter of his book to "silence" in premodern Europe. He used legal records, travelers' accounts, and etiquette books to analyze the principles underlying systems of silence. Many of his observations about preindustrial life mirror those of Schafer. Silence, Burke explains, was associated with showing respect or deference for those higher in the social hierarchy. Monks were supposed to be quiet in the presence of God, courtiers in the presence of the prince, women in the presence of men, children in the presence of adults, and servants in the presence of their masters (Burke 1993). In antiquity, Anne Carson claims, men were even supposed to control the sound of women in case women did not control their sound themselves. According to Sophocles "Silence is the *kosmos* [good order] of women" (quoted in Carson 1992: 127). Moreover, silence was, Burke argues, a sign of prudence, whether resulting from fear, from "the dissimulation of princes" or from "the discretion of the wise" (Burke 1993: 136). To these conceptions of silence, Foucault has added the notion of the *authoritarian* silence that disciplines the marginal Other to disclose his or her secrets (Said 1997: 16–20). And Smith (2001, 2003b) has found examples where nineteenth-century American slaveholders, distrustful of the ominous and resistant silence of their slaves, forced them into song and dance. Yet Burke focuses on silence as a sign of prudence in the presence of mighty others.

Burke suggests that premodern European systems of silence underwent changes between 1500 and 1800. An increasing concern with self-controlled speech arose during this period, which may have been related to the Reformation and its stress on ecclesiastical silence, as well as to the fear of spies at the courts of absolute monarchies. Burke speculates that "the bridling of the tongue" could even have been connected to the rise of capitalism, since a sixteenth-century Englishman made a comparison between "spending and saving words and money."[8] According to Burke, the rise of controlled speech was a general European trend, but restrictions were more effective in Protestant than in Catholic regions, thereby "widening the gap" between a "more silent" north and a more "talkative" south (Burke 1993: 140–141).

A connection that Burke does not mention, but is possibly relevant, however, is the rise of new forms of science, and the relation between modern science and sight. According to Bailey, "pre-modern societies were predominantly phonocentric, privileging sound over the other senses in a world of mostly oral-aural communication," whereas the "advent of typographic print in the fifteenth century gave a dramatic new saliency to visual perception" (Bailey 1996: 55). Apart from the disputed assumption that the rise of one sense necessarily means the decline of another, critical observations of another kind have been added to Bailey's account. Constance Classen, for instance, argues that *oral* cultures—oral in terms of the dominant form of communication—are not always *aural* cultures. She gives examples of oral cultures that symbolically orient themselves by temperature, by smell, or by color. Moreover, Classen claims that the standard ranking of the senses in Western culture, in which sight occupies "the highest position, followed by hearing, smell, taste and touch," originated from Aristotle, and thus preceded print culture by many centuries (Classen 1993; Classen 1997: 402). To complicate the story even more, the Greek tradition of conceiving truth and knowing as seeing was matched, as philosophers Hannah Arendt and Don Idhe underline, by a Hebrew tradition in which truth, like God, is audible but not visible (Ahrendt 1980/1977; Idhe 1976). Therefore, many academics simply refuse to consider the eye as *the* modern and the ear as *the* pre- or antimodern sensory organ (Erlmann 2004; Woolf 2004).

Yet many historians and social scientists support the notion that in Western society, knowledge production is closely associated with sight (Urry 2000; Smith 2000). According to the economist Jacques Attali, science has always "tried to look upon the world," desiring to "monitor" its meaning (Attali 1985: 3). Such statements should be qualified by saying that the profound changes that took place in natural philosophy in the sixteenth and seventeenth centuries, and the new reliance on observation and experiment as legitimate sources of knowledge, as well as the growing importance of print as a vehicle of communication, gave rise to a situation in which our scientific knowledge of the world increasingly came to be expressed in visual terms. Moreover, within the culture of seventeenth-century science, at least in England, the ancient topos of science and solitude was employed anew by stressing the importance of the "isolated individual in direct contact with reality" for the authentic discovery—not legitimation—of truth (Shapin 1991: 209). Although solitude does not run parallel with noiselessness, the emphasis on solitude suggests that silence too was likely to have been highly valued. Thus, as indirect as this connection may be, the rise of new forms of science in early modern Europe may have contributed to the increase in elite engagement in maintaining silence as Burke describes.

As James Johnson claims for Paris, however, the growing concern with maintaining silence did not stop in 1800. In late eighteenth- and nineteenth-century concert culture, for instance, it gradually became the norm to show deep respect, and thus silence, for the genius of music. This was the result of both musical and social changes, such as the introduction of absolute music,[9] which aimed at stirring authentic emotions within listeners, and the rise of the bourgeoisie (Johnson 1995).

In sum, the historiography and anthropology of sound make clear that noise and silence refer to deeply rooted cultural hierarchies. The right to make noise as well as the right to decide which sounds are allowed or forbidden has long been the privilege of the powerful, whereas those lower in rank (women, children, servants) were supposed to keep silent, or were under suspicion of intentionally disturbing societal order by making noise. Positively evaluated loud and rhythmic sounds have had connotations of strength, significance, and being in control, whereas noise as unwanted sound has often been associated with social disruption. Yet between the sixteenth and nineteenth centuries, the elite became increasingly obsessed with mastering its own sound.

Subsequent chapters show how the protagonists fighting against various noise problems dramatized sound by echoing, orchestrating, and transcribing the cultural connotations of noise and silence mentioned above. In their efforts, they also made use of auditory topoi of technology. What these topoi are about is the topic of the section below.

AUDITORY TOPOI OF TECHNOLOGY

In this section I will analyze a sample from a glossary of 1,084 quotations about sound from literary sources filed in the archives of the World Soundscape Project (WSP). As discussed in chapter 1, the historical part of the WSP project focused on documenting changes in the world's sonic environment over time. One way of doing this was to collect quotations about sound from literary sources, particularly novels.[10] In order to obtain an overview of the types of sounds mentioned during specific periods of time, WSP members classified all quotations into sound categories. These included natural sounds, human sounds, sounds and society, mechanical sounds, sounds as indicators, miscellaneous sounds, and sound experiences, quiet and silence, echo and reverberation, reporter's attitude to sound, mythological sounds, sound associations, and sounds relating to specific times of the day or year.

For the purpose of this chapter, several sets that include references to urban, mechanical, and industrial sounds have been selected from these categories.[11] Although urban and industrial sound should not in principle be blended with mechanical sound, the urban and industrial categories of sound have been included to avoid the omission of any relevant quotations. Of the sets studied, I have used only quotations published in Europe, Canada, and the United States between 1875 and 1975, nearly one hundred in total. Although the time span covered in the selection was 100 years, most of the quotations used below come from the years prior to and concurrent with the rise of the establishment of the first noise abatement societies in the early twentieth century.

Unlike the WSP members, however, I have not studied the quotations to give readers an indication of the sounds heard in the past. Such a descriptive approach carries with it the problems discussed in chapter 1, not to mention the methodological dangers related to using novels as direct windows through which one can listen to the sonic environment of the past. In contrast, I have analyzed the quotations with regard to recurring topoi, or stereotypical expressions, of mechanical sounds and the ways in which such topoi create drama.

When embarking on this analysis, I knew from the historical and anthropological literature on noise that the rhythm of the sounds described as well as their loudness and quantity could be relevant categories for analyzing ways of dramatizing sound. Don Idhe's *Listening and Voice: A Phenomenology of Sound* (1976) also made me aware of the potential significance of motion and spatiality. Idhe refers to recent research in biology and physics that suggests that humans use their ears to identify the character, location and direction of things, although in less advanced manners than animals. Idhe's aim in pointing out the findings of this scientific research is to substantiate his claim that philosophers have amply discussed the temporality of the auditory experience—sounds pass away—but unjustly neglected its spatiality. The anthropologist Steven Feld similarly stresses the significance of emplacement in his study of the "acoustemology" of the Kaluli in Papua New Guinea. In the rainforest, with its "lack of visual cues," the Kaluli know the time of day, the season of year, and their position in space through sound. (Feld 2003: 226–227). Such statements inspired me to take distance and direction, as well as rhythm and quantity, into account when analyzing quotations involving expressions of mechanical sounds. How does the structure of such expressions—their display of the quantity, distance, direction, and rhythm of sounds—helps to *stage* the drama of particular sonic sensibilities?

My analysis draws on, but also partially departs from, current approaches in literary studies. Recently, literary scholars have published several books and articles about sound. Some center on textualized sound in experimental poetics, poet-performances, and poetry (Morris 1997). Others focus on the expressions of sound in Shakespearean theater (Folkerth 2002) and twentieth-century plays (Meszaros 2005), or, like this section, on textualizations of sound in late-nineteenth and twentieth-century literary texts. For instance, Michael Cowan (2006) has written wonderfully on Rainer Maria Rilke's resistance to worldly noise in *The Notebooks of Malte Laurids Brigge* [*Die Aufzeichnungen des Malte Laurids Brigge*] (1910). And Philipp Schweighauser (2006) intelligently unravels the noises of naturalist, modernist, and postmodernist American literature.

Both Cowan and Schweighauser show how literature *represents* noise. Cowan for instance notes how Rilke's work is full of images of the artist in "search for a contemplative refuge from the noise of the industrial world," a topos frequently employed by the intellectuals of Rilke's time. In *Malte,* noise is represented as sound that malevolently interferes with concentrated, distinguished, and authentic thought (Cowan 2006: 126). Like Cowan, Schweighauser is interested in the literary representation of noise, but his "literary acoustics" also aims to analyze literature as a site for the cultural *production* of noise. By this he means that literature can be the white noise of culture by disturbing the main stream signals of society through an aesthetics of negativity and interruption (Schweighauser 2006: 197). Literature can give voice to noise, whether marginal, illusive, or inaccessible, without effacing it, often by employing new literary strategies. For instance, Stephen Crane vividly conveys the noisy violence of war in *The Red Badge of Courage* (1895) by exposing his readers to "chaotic fragments of subjective experience rather than a coherent plot structure" (Schweighauser 2006: 195).

I will refer to Cowan's work where he has an ear for the spatial expressions of sound in Rilke's *Malte,* and to Schweighauser's study, where he elaborates on literary techniques for communicating noise. My focus, however, is not on noise in general, but on mechanical noise. In addition, it is not the literary production of noise that interests me, but the conventional textualizations of mechanical sound in terms of its sonic characteristics. How do these characteristics contribute to the dramatization of the sound of technology? I have distinguished four ideal types of dramatized sound by classifying the quotations on the basis of the *quantity* of sound-sources described, the *distance* portrayed between the protagonist or narrator and the sound, the *direction* of the sound, its *rhythm,* and its *evaluation.* I have called the four ideal types the *intrusive* sound, *sensational* sound, *comforting* sound, and *sinister* sound, which are summarized in figure 2.1 below. Since some of the quotations were not informative enough, not all of them

Table 2.1 Auditory topoi of technology (Western novels, 1875–1975)

	Evaluation of sound			
	Negative		Positive	
	Intrusive	Sinister	Sensational	Comforting
Quantity	Multitude	Single	Multitude	Single
Distance	Close	Diverse	Diverse	Far
Direction	→ subject	Unclear	→ and ←	Unspecified
Rhythm	Irregular or unpredictable	Unspecified	Regular	Regular or unspecified

could be classified into one of the four ideal types. Furthermore, some quotations could be classified into more than one category. In most cases, however, it was clear to which of the four categories of ideal types they belonged. Below, I mention both the original source of the quotations and the card numbers of the WSP Glossary of Sounds in Literature (GSL) from which I obtained them. (Table 2.1.)

Intrusive sounds are usually expressed as a multitude of different sounds or a series of recurrent sounds. These sounds invade or threaten the existence of something or someone that is vulnerable or fragile, such as nature, harmony, or one's heart, mind, body or security. The noise frightens the protagonist and seems to move closer and closer toward or even *into* it, as in Romain Rolland's *Jean-Christophe* (1910): "The oaths of the drivers, the horns and bells of the trams, made a deafening noise. The roar, the clamor, the smell of it all, struck fearfully on the mind and heart of Christophe" (WSP GSL 845; Rolland 1910/ 1904–1912: 5). It is from the words "deafening" and "fearfully" that we know Christophe does not feel at ease with what he hears. Yet it is the multitude of the sounds heard that helps to create the drama.

The actual distance between the source of sound and the principal character, as in the case of aircraft, can, at times, be substantial, even though its sonic presence is near. Intrusive sounds violently enter the protagonists' world, often all at once, and endanger something cared for. The simultaneousness of different sounds creates chaos or forms a deep rumble that affects a subject's body. The following quotation from Virginia Woolf's *Mrs. Dalloway* (1925) expresses both the simultaneity and the unpredictability of the intrusive sound. "Suddenly Mrs. Coates looked up into the sky. The sound of an aeroplane bored ominously into the ears of the crowd.... Then suddenly, as a train comes out of a tunnel, the aeroplane rushed out of the clouds again, the sound boring into the ears of all people in the Mall, in the Green Park, in Piccadilly" (WSP GSL 205; Woolf 1925: 29–30). The narrator in Erich Remarque's *All quiet on the Western front* (1929) similarly depicts the invasiveness and unpredictability of noise when he explains that he has been "startled a couple of times in the street by the screaming of the tram-cars, which resembles the shriek of a shell coming straight for one" (WSP GSL 254; Remarque 1929: 167).

The quotations commonly present the intrusive sound in the context of urban surroundings, outside the subjects' homes. But there are also examples of the intrusive sound described in pastoral settings, as the following quotation exemplifies. "There was an awful sound. It was a train in the cutting. The hoarse panting echoed across the rise of the hill, through the trees, mounting up to me in successive waves of nauseous liberation. It shattered the wood with its steady reverberations, gasping and wheezing in its hidden trough. It seemed as if the monster would never reveal itself but simply menace with its struggle" (WSP GSL 64; Storey 1961/1960: 57). The intrusive sound of this train has "steady reverberations," but the sound is unpredictable at the same time: it pushes itself toward the narrator, but never reveals itself. When analyzing *Malte,* Cowan calls the "inability to fix the objects ... in space" an "imaginary animism of noisy things." The objects "*seem* to be coming toward him, although the very obstinacy of Malte's efforts to determine their spatial location betrays his own lack of mastery over his environment." The sounds "conspire in a deliberate violation of the subject's bodily boundaries," malevolent and external to the narrator (Cowan 2006, p. 139).

Even a gramophone's recording of a trumpet can be depicted as possessing animism and producing violent effects, as is the case Hermann Hesse's *Steppenwolf* (1921). "And in fact, to my indescribable astonishment and horror, the devilish tin trumpet spat out, without more ado, a mixture of bronchial slime and chewed rubber; that noise that owners of gramophones and radios have agreed to call music. And behind the slime and the croaking there was, sure enough, like an old master beneath a layer of dirt, the noble outline of that divine music" (WSP GSL 172; Hesse 1969/1921: 241). In this quotation, an age-old opposition between infernal, diabolical din and divine music or heavenly quiet comes to mind. In German and Dutch, expressions that convey one or both sides of this opposition, such as *Höllenlärm* and *Himmlische Ruhe,* or a *"laweit gelijk 'en oordeel, 'en hel"* (Heinsius 1916: 1192) date back many centuries.

Sensational sound is the positive counterpoint of the intrusive sound. Like intrusive sound, sensational sound refers to a multitude of sounds. Unlike intrusive sound, however, the sources of sensational sound can be felt both close and rather far from the narrator or protagonist, and, in all cases, fill the environment and surround the subject. Such sources are the crowds of the city, the movements of traffic, and the running of machines, and the sound may be directed either toward or away from the narrator. Here is Romain's *Christophe* once again, apparently in a different mood this time: "And all about him was the perpetual hum of Paris, the roar of the carriages, the surging sea of footsteps, the familiar street-cries, the gay distant whistle of a china-mender, a navvy's hammer ringing out on the cobblestones, the noble music of a fountain—all the fevered golden trappings of the Parisian dream" (WSP GSL 847; Rolland 1910/1904–1912: 203).

Moreover, sensational sound often expresses a regular rhythm, in complete contrast to the irregularity or unpredictability of intrusive sound. In other cases, a rhythm materializes by means of the sound's iconicity, such as the use of short sentences, parallelisms and enumerations, which are linguistic elements in a rhythmic pattern. Again, this contrasts with the literary strategies for communicating noise mentioned by Schweighauser, who outlines "fragmented" structures, "complexity of formal organization," and "multiple indeterminacies" in plots as the techniques for portraying noise (2006: p. 196). In the characterization of sensational sound, however, it is life, movement, energy, and power that predominate. In an enchanting and stunning way, these sounds are deafening. The narrator or protagonist often seems overwhelmed by a kind of urban sublime: "Turn your eyes upward, myriads of windows and balconies, curtains swinging in the sun, and leaves and flowers and among them, people, just to confirm your illusion. Cries, screams, whip-cracks deafen you, the light blinds you, your brain begins to feel dizzy and gulp air. You feel drawn into becoming part of the enthusiastic demonstration, to applaud ... but for what? ... It is Naples' life in its perfect normality, nothing more" (WSP GSL 414; Renato Fucini [1913], quoted in Valsecchi 1971: 182).

In other examples, the protagonist or narrator seems to be gripped by a more general animation, activity, or strength that the sounds of machines express. In *The Red Room* (1886–1887), August Strindberg makes the narrator hear "the clamour of the newly awakened town" rising "far below him": "down in the harbour the steam cranes whirred, the bars rattled in the iron-weighing machine, the lock-keepers' whistle shrilled, the steamers at the quayside steamed; the ... omnibuses rattled over the cobblestones; hue and cry in the fishmarket, sails and flags fluttering in the water ... and all this gave an impression of life and movement" (WSP GSL 933; Strindberg 1967/1886–1887: 2). And in Maxim Gorky's *The Artamonovs* (1925), Artamonov senior notes the "summoning blast of the mill whistle. Half an hour later would commence the indefatigable murmur and rustle, the accustomed, dull, but powerful din of labour" (WSP GSL 437; Gorky 1952/1925: 404).

In one quotation, a sequence and mix of intrusiveness and sensation highlights changes and ambivalence in the attitude toward modern life. "At first the Brangwens were astonished by all this commotion around them. The building of the canal across their land made them strangers in their own place, this raw bank of earth shutting them off disconcerted them. As they worked in the fields, from beyond the now familiar embankment came the rhythmic run of the winding engines, startling at first, but afterwards a narcotic to the brain. The shrill whistle of the trains re-echoed through the heart, with fearsome pleasure, announcing the far-off come near and imminent" (WSP GSL 78; Lawrence 1915: 7).

The *comforting* sound has, like sensational sound, a positive tone. Yet unlike sensational sound, a comforting one is a single source of sound, or a single type of source. It is often a sound heard at a certain distance, usually from an unspecified direction, but even if it is described as nearby, it does not intrude into the narrator's or protagonist's subjective experience. Its rhythm may be regular. Yet more common are descriptions in which the comforting sound is a continuous, soft drone of a nearby sound, without a clearly specified rhythm. In all cases the shelter, security, and harmony of the narrator or protagonist's direct environment (the bed, the house) are highlighted, with almost womblike associations, such as feeling safe and dry at home on a rainy day. The surrounding environment is also often quiet. At times, the distant sound can be the sound of an "indeterminate," yet positive "expectation," for instance when it is the sound of a train or carriage. According to Peter Peters (1998), the sound of the postal horn often conveyed this kind of meaning in nineteenth-century poetry and music. In these cases, the sound combines aspects of the comforting and the sensational topoi. Often, however, the auditory topos of the comforting sound is the urban and auditory version of pastoral and visual depictions of the Arcadian mill. It is a sole and consoling sound, the sound that underlines tranquility.

In Strindberg's *The Red Room,* the sounds of distant bells express a comforting quality at the end of the day: "And as, one after another, [the bells] fell silent, the last one could still be heard in the distance, singing its peaceful evensong" (WSP GSL 934; Strindberg 1967/1886–1887: 3). Bells are presented in a similar vein by Virginia Woolf: "As they looked the whole world became perfectly silent, and a flight of gulls crossed the sky, first one gull leading, then another, and in this extraordinary silence and peace, in this pallor, in this purity, bells struck eleven times, the sound fading up there among the gulls" (WSP GSL 206; Woolf 1925: 30).

Thomas Mann evokes the sound of luxury hotels in *Death in Venice* (1913): "A solemn stillness reigned here, such as it is the ambition of all large hotels to achieve. The waiters moved on noiseless feet. A rattling of tea-things, a whispered word—and no other sounds" (WSP GSL 210; Mann 1936/1913: 399). In *Doctor Zhivago,* Boris Pasternak situates another set of comforting sounds, again in a hotel: "It was warm in the hotel lobby. Behind the cloakroom counter the porter dozed, lulled by the hum of the ventilator, the roar of the blazing stove, and the whistle of the boiling samovar, to be awakened occasionally by one of his own snores" (WSP GSL 426; Pasternak 1958: 58). While sensational sound awakens the protagonist and lifts him or her up, the comforting sound makes him or her drowsy. "The workshop was quiet. The window looking out on the street was open. Lara heard the rattle of a droshki in the distance turn into a smooth glide as the wheels left the cobbles for the groove of a trolley track. 'I'll sleep a bit more,' she thought. The rumble of the town was like a lullaby and made her sleepy" (WSP GSL 424; Pasternak 1958: 25). It is the sound that accompanies the tranquility that makes it undisturbing, precisely because it is not completely silent: "The noise of the sixteen birchwood fires blazing in sixteen porcelain stoves made a pleasant break in the emptiness of the place" (WSP GSL 935; Strindberg 1967/1886–1887: 6).

And finally, the *sinister* sound is usually a single sound within a more or less silent environment. It functions as an ominous prognostication of what is going to happen. Its distance may be close or indeterminate, and its direction is often unclear. Its exact source and location are, unlike the comforting sound, not always evident, and it has no specific rhythm. It is the sound of suspense, like a creaky door in a haunted house. "It was the first time I had stayed in Grandpa's house. The floorboards had squeaked like mice when I climbed into the bed, and the mice between the walls had creaked like wood as though another visitor was walking on them. 'Is there anything the matter, grandpa?' I asked. . . . He stared at me mildly. Then he blew down his pipe scattering the sparks and making a high, wet-dog-whistle of the stem, and shouted: 'Ask no questions'" (WSP GSL 825; Thomas 1955/1940: 29–30). The following description of both human and nonhuman sounds is especially eerie: "As the ground floor split open and the debris thundered to the basement, one terrible cry came up from the shelterers beneath. Then silence, an aching empty silence broken only by small sounds; the rustle of broken water-pipes, the slow trickle of plaster-dust, a faint whimpering as if a child had bad dreams" (WSP GSL 361; Collier 1959: 123). Even the soft sound of electricity can create a shiver in a particular setting: "The narrow window opposite the bed had a blue neon sign hanging outside it; the sign flashed on and off, making an ominous buzzing noise" (WSP GSL 1022; Atwood 1969: 251).

The variety of sounds categorized as belonging to one topos or another suggests that stereotypical expressions of sound in literature are more or less independent from the source of the sound. The sinister sound, for instance, refers to human and nonhuman, and mechanical and nonmechanical sources of sound. At the same time, however, the examples make clear that in dramatizing the sound of technology, the quotations articulate particular aspects of mechanical sound. In the topoi of both sensational and intrusive sound, the multiplicity, movement, and continuity of the sounds of traffic and machines increase the tension. Although the single scream of a tramcar or whip-crack can be striking, it is the recurring character of the phenomenon and the mix of all sounds together that heighten the drama in particular. In the case of sensational sounds, it is the collective and ongoing rhythm that contributes to the textual "staging" of their positive effects. In the case of intrusive sounds, it is the irregularity, chaos, and unpredictability of sound—the opposite of cadence and rhythm—that produces part of the dramatic performance.

The presentation of distance and direction also do additional dramatic work. The continuity of the comforting sound creates a soothing hum, but its distance and unspecified direction ensure its innocence. In contrast, the movement of sounds toward the narrator helps to stylize sound as intrusive noise, whereas a lack of knowledge of the exact location and source of sound contributes to its dramatization as a malevolent or eerie sound. Mechanical repetition, however, is responsible for the inexorability *and* the untiring animation of sound, and, in referring to the continuous character of mechanical sounds, helps conjure up the soothing spirit of the comforting sound, the dynamics of sensational sound, and the inescapability of the intrusive sound. That is why I have chosen to speak about "auditory topoi of *technology.*"

CONCLUSIONS: THE LONG SHRIEK

In *The Machine in the Garden,* Leo Marx describes how, in nineteenth-century-American literature, "the long shriek" of the locomotive symbolized the disturbance of the pastoral ideal. The noise aroused "a sense of dislocation, conflict and anxiety." It referred to the tension between industrialization and the rural, green, orderly, and quiet environment, or the ideal of withdrawal "from civilization's growing power and complexity." Just like Mark Smith has done, Leo Marx claims that "the rhetoric of the technological sublime," with energy, power, and grandeur attributed to the machine, dominated American magazines of "the governmental, business and professional elites" (Marx 1964: 16, 9, 219). Yet literary writers positioned the rattle of railroad cars and the thunder of engines as the signs and symbols of dissonance. As in the literary quotations above that expressed the intrusive sound, the writers discussed by Leo Marx dramatized the sounds of technology by presenting them as powers ruthlessly invading a realm to be defended. Again, evoking an oncoming direction, motion, and continuity of mechanical sound contributed to the drama envisioned.

This example suggests the relationship between the auditory topos of the intrusive sound, in this case, the shriek of the locomotive mixed with the rattle of railroad cars and the thunder of engines, and the symbolism of noise. By positioning, in the auditory topos of the intrusive sound, the chaos of train sounds against the order of the countryside—the situation to be protected— mechanical noise is the symbol of social disruption. Similarly, the auditory topos of sensational sound brings in the meaning of strength associated with positively evaluated loud and rhythmic sounds, notably when re-enacting repetition. And strikingly, the topos of the comforting sound resembles the pastoral theme in music, which, according to cultural geographer George Revill, sorts out "order from chaos and civilisation from barbarism" like the ringing of bells could signify in Corbin's description of the French countryside (Revill 2000: 600–601).

As subsequent chapters will illustrate, people complaining about noise often employed the topos of the intrusive sound. They included the topos's stress on the vulnerability of the world or body being invaded, and its grasp of sound as a discrete entity with a distinct, malevolent agency. Activists putting noise on the public agenda often employed the same dramatizing style. Their repertoires of dramatizing sound, however, did not always fit in with the kind of dramatizations required in the contexts in which they acted. Nuisance law hearings and shop floor culture, for instance, asked for rather different dramatizations of noise. The next chapter will show how this clash between dramatizations influenced the fate of public problems concerning industrial noise.

A Continuous Buzz:
The Ongoing Charge of Industrial Noise

The Murder of Sleep

In the September 1900 volume of the *Westminster Review,* the British landscape painter George Trobridge voiced his concerns over the effects of noise. For Trobridge, one of the "most alarming concomitants of modern civilisation" was "the increasing prevalence of lunacy and other nervous diseases." This was due to "the hurried, anxious lives that we lead." To those "utterly exhausted," undisturbed sleep was "the best of all tonics." Noise, however, murdered sleep. What followed was an account of "the increase of distracting noises in recent years." Trobridge complained about the "clink, clink" of the bricklayers' trowels, the "intolerable nuisance" of the traction engine, and the "hideous shrieks" of the steam tug's siren. He also found the sounds of electric tramcars, the steam tramway, and the "constant rumble of the traffic" disturbing (Trobridge 1900: 298–301).

Many people, Trobridge suggested, considered such noises unavoidable. "The march of civilisation brings inconveniences as well as benefits, and we must be content to take the thick and the thin together. Certainly! Certainly! But something *can* be done to lessen distracting noises at night." He proposed abolishing the "screeching" whistles of manufactories, contending that workmen would rather have "callers" wake them up. In addition, Trobridge wanted the horns of railway engines moderated, the use of traction engines during the night banned, the pavements of streets reconstructed to make them quieter, and "night rowdyism" declared illegal (Trobridge 1900: 302).

In his article, Trobridge referred to industrial noise in two ways. He complained about the disturbing noises during the night that came "from the glass-bottle works, lately erected just across the way, and in operation day and night." The problem had arisen, he explained, because "part of the work is carried on in an open shed, and the night-workers amuse themselves by singing and shouting at their work" (Trobridge 1900: 299–300). Yet Trobridge also described how he "drop[ped] asleep to the sweet music of a tilt-hammer and the rushing of steam at a distant iron forge, to be roused again, perhaps, about four in the morning, by an early excursion train" (Trobridge 1900: 301). Trobridge portrayed the sounds he heard at the glass-bottle works—the chatter and singing of the laborers—as close and continuous, in a manner that is in line with the auditory topos of the intrusive sound. He described those of the iron forge, however, as distant and musical, a mechanical lullaby that helped him fall asleep. Here, Trobridge used the topos of the comforting sound just before describing the intrusive sound of the early train, thereby heightening the drama of the train's sudden intrusion into his valuable and vulnerable sleep. Intriguingly, the iron forge's comforting sound is in line with Trobridge's message. It was not the sounds of industry at large that caused so much disturbance, but the *unnecessary* ones at night and in the early morning.

In Trobridge's rendering of the issue, the dramatization of industrial sound and the delineation of the noise problem were perfectly in line with each other. This chapter, however, takes up various examples of mismatches between the dramatization of industrial sound and the definition of industrial noise in particular settings, resulting in clashes that contributed to the persistence of noise on the public agenda. The first case involves nuisance law hearings in which citizens were allowed to express their objections against the sound of particular industries. The way in which they dramatized their complaints, however, did not fit in with the framing of industrial noise required by the nuisance legislation. The second case is about factory workshops where physicians attempted to persuade workers of the risks of noise-induced hearing loss in the hopes that they would agree to wear earplugs. Their definition of the industrial noise problem did not match with the dramatization of mechanical sound by the workers, though. The third case unravels the discussion about playing music on factory shop floors to mask industrial noise. In this case, the definition and dramatization of industrial noise produced by experts, managers, and workers were largely in line with each other. Even this did not help to remove noise from the public agenda, though, since the shared approach resulted in the acceptance of background music that eventually came to contribute to noise problems beyond the realm of work.

The most long lasting interventions concerning industrial noise involved the establishment of islands of silence and the regionalization of industrial noise. Understanding how the clashes between context-bound dramatizations and definitions kept noise on the public agenda is one purpose of this chapter. Understanding how industrial noise came to be cast in spatial terms is the other. What has made the spatial solutions so important, and what have been their consequences?

NUISANCE LEGISLATION:
ISLANDS OF SILENCE AND THE MISPLACED COMPLAINT

Long before industrialization, the noise produced by various skilled trades was the object of many local ordinances, as were many other forms of noise. Local bylaws targeting singing and shouting on Sundays, barking dogs, crying vendors, nightly whistling, street music, and making noise in the vicinity of churches, hospitals, and other institutions were commonplace. Most of them prohibited making noise on particular days, at particular hours, in particular places, or without a license.[1] Local regulations that were drawn up to police the noise made by blacksmiths, coppersmiths, mills, and other trades were based on a similar approach: they placed restrictions on both when and where these industrial activities could take place.

The oldest prescriptions on the trades date back to ancient Rome. Roman law prohibited coppersmiths from establishing their trade in any street where a professor already resided. In 1617, the law faculties at the universities of Leipzig and Jena, consciously or not, imitated Rome by proclaiming that "no noisemaking handworker" was permitted to set up his business at any place inhabited by "Doktores." Both legitimated their actions by claiming that the handworkers' noise forced "the scholars to give up their research." More than a century later, in 1725, the university town of Turin banned the work of all noisy handworkers from the town's borders (Wiethaup 1966: 121). By that time, all the coppersmiths' workplaces in Venice had already been centered in one quarter (Rosen 1974: 514). This spatial intervention in the noise of one particular trade was not a Venetian invention. In 1394, an Antwerp ordinance ordered that citizens were not allowed to put up a grease or flatting mill within its inner walls (Poulussen 1987: 102).

The noise of mills was a common nuisance to city dwellers in early modern Europe. They literally hammered no less than smiths did. In Antwerp, there were grease mills for pressing oil from crushed seed, snuff or powder mills for grinding tobacco, and madder mills for milling and mashing madder, a seed plant used for the production of the color Turkish Red. In 1691, the municipal authorities decided that owners of grease mills must obtain a license before starting their business. This was, again, a regulation with spatial consequences, since granting or withdrawing licenses had a direct effect on the density of trades in particular areas within the city's walls. Five years later, the council began regulating the hours of production: grease and soap makers were required to restrict their seed crushing activities to the hours between 5:00 a.m. and 10:00 p.m. Authorities in Brussels instituted the same regulation, and Bergen officials adopted a similar one for iron forges in 1737. Apparently, the rules were difficult to uphold, because Antwerp citizens were still complaining about "the violent noise and big roar" in the vicinity of mills in the mid-eighteenth century, which led to the doubling of fines in 1765 (Poulussen 1987: 80, 107, 114, 160–163).

Despite the evidence of actions taken against trade noise, especially the hammering trades, several historians stress that until the late nineteenth century, urban dwellers and authorities worried less about noise than about nuisances such as stench, soot, smoke, and the risk of fire. According to Peter Poulussen, preindustrial citizens took action if the trades seriously endangered their goods, production, health, or their enjoyment of everyday life, but not in the case of a mere inconvenience (1987: 80, 124). Alain Corbin claims that in France, authorities and investigators still expressed such an attitude in the mid-nineteenth century, mentioning noise less often than other nuisances created by industry. The *police sanitaire* "even refused to recognize noise pollution." They and other authorities justified the legislative silence on this issue by the "traditional rowdiness of artisans" and the "cries of Paris": the commonplace cries of vendors. The only exceptions were the constraints to which "gold-beaters" and the owners of "heavy castings forges" were subjected (Corbin 1995/1991: 156). In the course of the nineteenth century, however, both the remarks and complaints about industrial noise increased, at least in France (Corbin 1995/1991: 156; Corbin 1986/1982: 173, 177; Massard 1999: 14).

Which laws, then, were of help to late nineteenth-century European citizens in cases in which they sought to lodge an official complaint about industrial noise? Many continental European countries had instituted regulations against the "disturbance" of the tranquility of neighborhoods at night, or, as in Germany, even during the day. The regulations were usually aimed at indecent sounds, offensive noise, and rude behavior—of drunks, for example. English common law employed a more general notion of "public nuisances" or "misdemeanors," which involved "a material interference with the comfort and convenience of life" of all persons nearby. Carrying on trades and activities "likely to cause a noisy crowd to gather" could be considered such an offense (Kerse 1975: 15–16).

Unlike the English notion of public nuisance, however, the Dutch, German, and French penal code excluded noise created by trades and factories. In France, a coppersmith working after 11 pm could not be punished unless a local ordinance fixed the hour that he must stop (van Dam 1888: 9–10). In the Netherlands, the notion of "disturbance" did not include the nightly sounds of bakers, butchers, and other trades, unless the noise was created by "mischief" (van Dam 1888: 40).[2] In Germany too, the penal code excluded the noise produced by trades. Here, noise could only be punished if it was found to be *aimed* at disturbing the peace of others (van Dam 1888: 30). Over time, the scope of the law in Germany was broadened to include the *unnecessary* noise of trades, such as the shortening of iron girders in open space at night. Yet, the legal options for abating industrial noise remained very limited (Saul 1996b: 171).

Civil codes also restricted the actions persons could take against those making noise. The Dutch civil code prohibited all deeds that involved damaging property or depriving persons from freely enjoying it (Tjaarda Mees 1881: 164–166). It was not until 1914, however, that the Dutch Supreme Court recognized noise as a cause of the deprivation of property, and it counted as such only when it occurred in the form of "terrible din, heavy drones and severe vibrations." A few years later, judges also accepted actions that clashed with good manners as a cause of the deprivation of property. Winning a case on noise by demonstrating that an action ran counter to societal decency, however, was far from easy.[3] Similarly, the German civil code was written in such a way that it only served those who were able to prove that the noise that bothered them interfered more than "only insignificantly" with the use of their property and exceeded "the level that was common in the area." "For upper-class residential areas or summer resorts," the socioeconomic historian Franz-Josef Brüggemeier claims, "this paragraph offered help. Here, it could be argued that it was not normal to have factories and that their presence reduced the value of properties." In the case of industrial areas, however, this kind of reasoning could not be upheld so easily. The civil code assumed that citizens who owned property in industrial areas profited from the increases in the price and rents of their property, as a consequence of its proximity to industry. Property owners were therefore expected to be willing to withstand the disadvantages of their property's location, even if they involved the annoyance of a steam hammer's shocks (Brüggemeier 1990: 221; Saul 1996b: 172).

Under English common law, the abatement of "private nuisances," or anything that interfered "with a person's use or enjoyment of land or of some right connected with it" focused on the protection of property against damage and, again, took noise for granted (Kerse 1975: 17). Damage was defined as the loss of a property's value or its habitability, or a substantial personal discomfort, and had to be permanent. Such damage, however, was difficult to prove in the case of noise. In an exemplary case, a landlord's claim against a nearby railway company, which he accused of having caused his tenant to leave, was denied: the loss of rent did not self-evidently lead to a permanent loss of property value (Kerse 1975: 22–23). Yet in another case, a complainant who took offense to the sound of two circular saws—"a continuous buzzing ending sometimes in a discordant shriek and sometimes in a melancholy wail"—in a residential area, *was* granted an injunction to prohibit the operation of the saws (Kerse 1975: 27). Apparently, this mechanical sound was considered exotic to the particular locality's character. Even in these cases, however, the sole option available to a complainant was to ask for compensation for the damage suffered (Tjaarda Mees 1881: 4; Kerse 1975: 26).

In all three countries, people without property had a very limited chance of winning a case against noise under civil codes or private nuisance regulations. Although English common law mentioned "occupiers," which implied both owners-occupants *and* tenants, the relatives living with such persons had no right to sue. From 1906 onward, German civil code allowed tenants to terminate their lease without notice in cases where noise could be shown to endanger their health. The machinery din of a butcher's shop on the ground floor of a residential house was one such occasion. By threatening to leave, tenants could force owners to take action against the noise. According to Saul, this clause proved far more helpful for well-to-do citizens who would be welcomed on a large market of spacious apartments than for laborers who usually had difficulties finding a cheap place to live (Saul 1996b: 174).

The third option available to citizens for standing up against industrial noise grew out of a series of national laws regulating the establishment of factories. In 1848, England created the Public Health Act, which gave local boards of health in particular regions the right to supervise annoying and health damaging factories and to make the right of their establishment dependent on permissions. The Health of Towns Act of 1853, the Nuisances Removal and Diseases

Prevention Act of 1855, the Local Government Act of 1858, and the Sanitary Act of 1866 all extended the rights of local authorities to supervise, inspect, and discipline factories for nuisances. Commands to remove nuisances were impossible without official complaints, however. The system was thus merely a *repressive* one (Tjaarda Mees 1881: 8–20). Moreover, under both the Local Government Act and the Public Health Act, complaints could be dismissed in cases in which the owner of a business or factory had adopted "the best known means" or the most "practicable" way for preventing a nuisance (Brimblecombe and Bowler 1990: 186, 191).

In Germany and the Netherlands, a *preventative* approach prevailed. The Dutch installed a Nuisance Act in 1875.[4] It provided citizens with the right to be informed about requests for permits to start, expand, or reorganize a particular group of industries, as well as the right to object to the granting of such permits. In cases in which citizens used this title to protest, local authorities were obliged to organize a public hearing to allow citizens to clarify why they objected to the action in question. Subsequently, professionals, such as the inspector of health, checked whether or not the citizens' complaints made sense (Prins 1991: 22, 30). In addition, the act allowed local authorities to allot particular areas to particular industries and to forbid their establishment in other areas (Tjaarda Mees 1881: 227). One of the rationales behind the law was to protect public health and safety. The issue at stake in the debates preceding the law's enactment, however, was the conservatives' wish to uphold the value of land and real estate against the encroachment of new industry versus the liberals' desires to safeguard it from narrow restrictions and unwarranted complaints (Diederiks and Jeurgens 1988: 433; Aalders 1984: 134). The result was that permits could be refused only in the case of danger, damage to property, to business, or to health, or a severe hindrance that made houses uninhabitable or hindered the regular use of institutions, such as schools, churches, and hospitals (Prins 1991: 22). In practice, many license procedures started only after neighbors had lodged a complaint, since many trades did not have the permission they required (Diederiks and Jeurgens 1988: 434).

The Dutch Nuisance Act closely resembled the Prussian *Allgemeine Gewerbe-Ordnung* (General Trade Code) that took effect for the whole of the newly constituted League of North Germany in 1871. Citizens running forges, boiler, tin, and iron factories or other industries involving riveting required a license before starting their business. Initially, a reasonable distance between the forges and residential housing sufficed for obtaining permission. From 1895 onward, however, the conditions for obtaining licenses became stricter. The forges had to reduce their putative nerve wracking noise by creating double walls, double doors, and double windows. What's more, the minimum distance between riveting and residential areas had to be one hundred meters and between riveting and the street, thirty meters.

At times, the German trade code overruled regulations enforced by local police that had been relatively tough on the noise of smiths (Saul 1996b: 165–167). In general, however, the *Gewerbeordnung* offered a few more possibilities for noise abatement than the Dutch Nuisance Act. Unlike in the Netherlands, in Germany, authorities could invalidate the licenses of industries that had already been established, provided that they agreed to compensate for the industries' losses (Tjaarda Mees 1881: 107–108). Moreover, even industries that did not have to apply for a license but were extraordinarily noisy could be ordered to report their activities. Their work could be forbidden or restricted in cases in which it severely disturbed churches, schools, hospitals, health institutions, or other public institutions. What all industrial nuisance laws in Europe shared, however, was the privilege of granting hearings in which the arguments of complaining citizens could be weighed against those accused of creating the nuisance. On such occasions, complainants had to stage their dramas in a way that matched the interpretations of the law to be convincing.

According to lawyers discussing the Dutch Nuisance Act, authorities could refuse a permit only to industries that were extremely noisy. It simply "did not do to exclude" all industries using machines "from town areas." Public institutions such as schools, churches, and hospitals deserved a higher degree of protection than private homes, however. The argument for schools and churches was that "the spoken word had to reach so many more people and over such a bigger distance than in ordinary conversations," and for hospitals it was "the presence of persons being extremely sensitive to noise" (van Ommeren and Hoendervanger 1914: 1–4, 114, 116; Cramer 1891: 26–28; Cramer 1904: 66–67). In other words, schools, churches, and hospitals should be *islands of silence*.

The idea that learning and contemplation required tranquility was far from new. The association between science and solitude, for instance, was rooted in part in the conviction that "separation from society was the most authentic life for a Christian," as sociologist of science Steven Shapin has shown. Solitude was the best setting for the "attainment of genuine religious knowledge" (Shapin 1991: 197). And although solitude should not be conflated with silence, a religious man like Thomas à Kempis *did* want to study the bible in solitude *and* in silence (Overmeer 1999: 185–186). During the Renaissance, "the role of the scholar as a relative isolate was increasingly acknowledged as a fact of social life. The monastery, the college, the closet or study, and even the laboratory, observatory, and garden, were places where the scholar could be found at work" (Shapin 1991: 198–199). These practices acquired new strength in seventeenth-century England where the importance of the "isolated individual in direct contact with reality" was increasingly stressed for the discovery of truth (Shapin 1991: 209).

The notion that the ill deserved tranquility had a similarly long history. In Antiquity, for instance, Galen promoted a balance between undisturbed sleep and activity for retaining one's health (Burns 1976). And Hippocrates recommended keeping shelters built for the sick away from noise. In many nineteenth-century European towns, citizens put straw or sand on the pavement in front of hospitals and the homes of the sick to reduce the roar of traffic (Verslag 1934: 21; Bailey 1996: 59; Schafer 1994/1977: 190; Cowan 2006: 132). Industrial nuisance legislation adopted such long-standing usages and beliefs by protecting hospitals, schools, and churches from noise. The cultural associations of contemplation, learning, and convalescence with silence and solitude, the need to communicate with large groups of students and worshippers, and presumably, the fact that churches, schools, and hospitals were easy to identify helped establish their exceptional position in nuisance legislation.

Consequently, people speaking on behalf of such institutions were less often overruled than individuals trying to convince authorities of the legitimacy of their complaints. In Germany, "it was only in spas or near hospitals that health risks could be cited to prevent the opening of new factories" (Brüggemeier 1990: 221). German schools were also protected, but only to a certain degree. In one case, a locksmith was not allowed to construct iron girders in a courtyard near a school, while in general making iron doors and garden fences in a courtyard was permitted (Saul 1996b: 165–166). In the Netherlands, institutions also had a relatively good chance of getting their claims against noise accepted by the authorities. The Dutch historian R.E.J. Prins analyzed 795 requests submitted to authorities in the Dutch city of Kampen for a permit to establish a trade or business between 1875 and 1940. He found that in 159 cases, citizens had lodged complaints against them, and in 44 of these noise was identified as an issue. In a similar project, the historian A.M.N. Verberne examined 4,594 requests submitted to authorities in Tilburg between 1875 and 1937. Of these, 369 contained complaints (Prins 1991: 109–110; Verberne 1993: 9).

This research reveals that the number of complaints lodged against permits was low compared to the number of permits requested. That many workshops had another shop next door may have been one of the causes. Lodging a complaint not only endangered one's neighbor's means of subsistence, but might also reduce the complainant's own chances of getting a permit, since his or her neighbor could come up with a counter-complaint (Prins 1991: 32; Verberne 1993: 50, 76; Metz 2002: 5). It is no coincidence that there were fewer complaints lodged in the German Ruhr area than in the Berlin upper class suburb of Dahlem. According to Brüggemeier, this was due to the chance of making money in a Ruhr area industry. The underrepresentation of the bourgeois did the rest, since they were the ones in Berlin and elsewhere that complained most frequently. (Brüggemeier 1990: 219; Saul 1996b: 162). According to Verberne, the number of complaints in Tilburg did not increase with economic activity, but it *did* increase with population growth. This finding suggests a role for population density, next to class and profession, in the origin of complaints (Verberne 1993: 73, 76).

Of all the objections against noise raised on behalf of institutions, only about half were successful, at least in the towns that Prins and Verberne examined. In 1906, for instance, the governors of Kampen's municipal hospital objected to the establishment of a stonemason's workshop near it because the "knocking of the hammers" would bother the patients. Local authorities agreed (Prins 1991: 70). A few years later they responded similarly to opposition against a request to install a planing-machine in a shed near the hospital's wing for tuberculosis patients. As the hospital governors claimed, the machine's dust and "piercing" sound would disturb the patients' recovery (Prins 1991: 71). In 1888, one of Tilburg's forges had been converted into a steam boilermaker's workshop. The vicar of a nearby protestant church lodged a complaint, although he felt "slightly embarrassed" about protesting against a fellow citizen's profession. The sound, however, was a severe nuisance to the divine service. The local authorities supported the vicar's protest and ordered the smith to reduce the nuisance (Verberne 1993: 50). Yet even churches did not always get their way. In 1896, the Dutch Council of State refused to deny a permit allowing a forge to be built near a church. A public street ran between the church and the forge, with many passing vehicles and trams, and the smith did not work on Sundays (Cramer 1904: 67). Intriguingly, public worship was not the only issue the authorities had taken into consideration. That the place had already *been* noisy had also been significant in their decision. Legitimizing new noise on the basis of existing noise was akin to the civil code practice in Germany. Usually, however, complainants speaking on behalf of institutions had a reasonable chance of success.

The chances of success were much smaller for individual citizens. Time and again their objections were overruled. Complainants described the noise made by animals, the moos, bleats, snores, and screams of cattle and sheep in slaughterhouses. They complained about the noises created by machines as well: the myriad mechanical sounds made by vibrating pumps, steam engines, gas engines, sawing machines, and weaving looms, machines installed in factories for the production of furniture, metal goods, and wool. The roar of the many forges—whose workforce still worked an average of between 60 and 70 hours a week at the end of the nineteenth century—was a constant subject of complaint (Prins 1991: 62, 67). So was the noise made at night by kneading machines in

bakeries, by stonemasons, boot makers, coopers, chandlers, distilleries, print-
ing offices, and, later, by service stations for cars. Yet, despite all the colorful
and detailed descriptions that individual complainants produced, in most cases,
local authorities granted requests for the establishment or extension of trades
anyway.

Many individuals complained about the loss of rent, and thus of the
income, they received from their tenants. The damage to their property caused
by vibrations was another type of problem they mentioned regularly (Prins 1991:
40; Verberne: 1993, 33, 36–37, 40–41, 51, 70; Diederiks and Jeurgens 1988: 178).
At times, such individuals tried to make clear that their house had lost its value
or their neighborhood its attractiveness as a result of noise (Verberne 1993: 53;
Metz 2002: 6). Given the law's formulation, these types of laments are not surprising. Less self-evident, though, are the complainants' frequent references to
their ill health, bad nerves, and hard living conditions, as well as to the frailness
and personal troubles of their wives, sisters, and parents (Prins 1991: 40, 61, 70–
71; Verberne 1993: 40, 48, 49, 52, 58, 69–70). In most of these accounts, people
stressed their vulnerability or that of their spouses. One woman who objected to
the establishment of a steam flour mill not only complained about the expected
loss of the rent she would receive on the three houses she owned near it, but
added that she was "a widow indeed" (Prins 1991: 40). One complainant was
said to be haunted by a "weakness in the head" and another spoke on behalf of
a father, who had to cope with a noisy forge near his bedroom while having
"reached the age of eighty-four" (Prins 1991: 61). The director of a telegraph
service deplored the noise that disturbed the sleep of his "wife, already suffering
from nerves" (Verberne 1993: 40). Another man stressed that his wife, at "the
age of seventy-three," was so troubled by noise that she felt forced to leave their
home occasionally "so as to calm down her nerves." She cried frequently, he
explained, since "we have been so deceived" (Verberne 1993: 49).

It could be argued that citizens emphasized their fragility because the law
protected hospitals and stipulated damage to health as a legitimate reason for
denying requests for permits to establish a trade. Yet lawyers stressed that "objections of an extraordinary kind, such as the accidental illness of some inhabitants"
would *not* be relevant (Tjaarda Mees 1881: 161). Despite their warnings, this was

———

exactly the type of argument that individuals tended to use when lodging a complaint. They described their health and nerves as already being in poor condition, or predicted how a particular noise would undermine their health in the future. What they did, in fact, was to heighten the drama of their complaint by employing the auditory topos of the intrusive sound. Since complainants probably knew that it was difficult to prove that noise made their houses uninhabitable or caused damage to their health and property, they depicted their weakness and that of their relatives in great detail. Some citizens, like the man whose wife had to flee her home to calm down, employed what sociologists have termed "members' measurement systems." In these cases, complainants explained how noise had "forced them to make changes to their everyday activities" (Burningham 1998: 543). In the context of the Nuisance Act, however, these ways of formulating complaints were tragically *misplaced*.

In line with the characteristics of the intrusive sound discussed in the previous chapter, complainants stressed the recurrent nature or even infinity of the mechanical sounds they despised. They spoke about "incessant knocking," "continuous tumult," "constant hubbub," and "everlasting hum and drone" (Prins 1991: 67, 62, 61; Metz 2002: 7). In addition, they stressed the aggressiveness and bodily dominance of the sounds by mentioning the "drone," "violence," "shrieking," and "deafening din" of the machines (Prins 1991: 86; Verberne 1993: 13, 70). According to a neighbor of a flour mill that used small carts with iron wheels on wooden floors for internal transport, the racket produced was like a "cavalcade of jumping kangaroos" (Verberne 1993, 36). He thus dramatized the chaos of the sounds and attributed a malevolent animism to them. Also, by referring to the thin partition walls between their houses and those of the noisemakers—something many complainants did—the closeness of the sounds came to be highlighted (Prins 1991: 42, 46, 69; Verberne 1993: 52).

A complaint lodged by a woman against her neighboring stonemason in 1899 illustrates the most common way of bemoaning one's lot: "that the drumming, hammering and ticking, produced by the . . . stonemason during the entire day, is heard by her and her inmates every day from morning till night and even assumes such enormous proportions that her house, although very soundly constructed, rumbles and the objects in the living room, although separated from the stone-mason by a broad passage, rattle: that all this is of great harm to her house that trembles continuously by such drums and drones; that finally her health will obviously suffer and she already experiences that her nerves get damaged—not to speak about the tragic consequences that would follow from the noise in case of illnesses" (quoted in Prins 1991: 69).

Notwithstanding the talent for drama many citizens displayed, local authorities often overruled their complaints, and rarely refused requests for permits. In fact, Kampen citizens proved successful in only 15 out of 434 requests submitted between 1904 and 1933 (Prins 1991: 87). Even a group of complainants who objected to a request to construct a fourth forge in a neighborhood proved unsuccessful in preventing its establishment. The law only enabled complainants to take action against *individual* trades (Prins 1991: 64). The result was, once again, comparable to the effect of the German civil code practice that made citizens' complaints in quiet neighborhoods more effective than the charges made by inhabitants of noisy areas.

Many of the hammering crafts that had caused serious noise nuisance, such as the cooperages and copper, tin, lead, and zinc smiths, gradually disappeared by the end of the nineteenth century and after, or were confronted with more rigorous restrictions to their working hours (Prins 1991: 65–66, 62, 70). Around the same time, gas and oil engines replaced steam engines, and a couple of decades later, smaller electric motors superseded them. According to Prins, these changes reduced the noise of individual machines (1991: 93–94). Verberne, however, claims that mechanization undid the initial improvements. And although the electric motors were relatively quiet, complaints about them increased (Verberne 1993: 82). In 1929, Amsterdam experienced a rise in complaints about these motors, which were often employed in simply housed industries, that was large enough to prompt city authorities to create a special bylaw to grant permits for the use of electric motors that were not covered in the Nuisance Act (Tjaden 1929: 190). Two additional changes help to explain the increase in complaints despite the reduction in the number of forges and of noise *per* machine. One was the introduction of the radio. Electric engines, so complainants noted, interfered with the reception of the broadcasts they wanted to listen to on the radio (Verberne 1993: 44, 52). The other relevant change was the increasing use of telephones in offices. The din of neighboring industries disturbed telephone calls (Verberne 1993: 48). The Nuisance Act could not have been of help to citizens who complained about such problems, however. Even apart from the exemption of small engines, not being able to listen to the radio or talk on the telephone was not included in the definition of damage to health and property or of condemned housing.

Clearly, the repertoires of dramatizing sound employed by individual complainants of industrial noise usually clashed with the rationality behind the law: these dramatizations were thus tragically out of place. The trade codes provided citizens without property with opportunities to object to industrial noise. The idea behind this legislation, however, was not only to protect a generally defined public health from the downsides of industries, but also to balance conservatives' fears about the loss of value of their tenements against liberals' desire to expand their industries. Therefore, the key to a successful complaint was to demonstrate *damage* rather than one's *vulnerability,* as many complainants in the Netherlands did by invoking the topos of the intrusive sound. As a consequence of the islands of silence tradition, churches, schools, and hospitals had a slightly better chance in finding protection from noise.

It was this idea of regionalizing noise by creating zones of industrial activity and zones of tranquility that became the most important legacy of the nuisance legislation. The idea of zoning itself was the heir of a few earlier practices, such as separating the hammering trades from the learned professions and muffling the din of traffic in the proximity of the sick. Before industrialization, much labor had been done at home. During the first phases of industrialization, it was still considered efficient to build factories in the vicinity of the labor classes' public housing. In the course of time, however, industrial activities gradually acquired their own, more enclosed, spaces (van Daalen 1987: 50, 67). Environmental historians have called this process the "regionalization" of "pollution" (Brüggemeier 1990: 214). Solutions to noise in terms of space were, as such, not new, but they became dominant at the expense of restrictions in time—one of the seventeenth- and eighteenth-century solutions to the noise nuisance caused by grease and flattening mills. Spatial solutions seemed like the logical middle ground between a conservative policy of land and property protection, a bourgeois policy of industrial expansion in time and space, and a cultural legacy of keeping the places of learning, religion, and recovery as quiet as possible. It was not that restrictions to making noise in terms of time vanished altogether. In Germany, for instance, local ordinances kept protecting citizens' night rest.[5] Yet, the noise of trades became increasingly cast in spatial terms. After the Second World War, local authorities in the Netherlands stimulated industries to leave the city center and to settle down in particular factory areas (Prins 1991: 8; Metz 2002: 8). Planning control acts, introduced in many European countries from the 1960s onward, helped increasingly to structure this process (Kerse 1975: 56–66; Wiethaup 1967: 74).

Enclosed spaces for industrial activities evidently reduced the presence of industrial noise in residential areas. Yet, tragically again, it raised the need to commute, and thus the noise of traffic (Amphoux 1994: 86). This, in turn, inspired interventions that set maximum levels of noise for those areas in which residential housing and busy roads mixed. In England, for example, the Control of Pollution Act of 1974 provided local authorities with the right to designate "noise abatement zones" in which prevailing noise emission levels had to remain steady or be reduced (Kerse 1975: 49–55; Garner 1975). Without the means to *measure* and to *rate* noise, which will be discussed in subsequent chapters, zones with maximum levels of noise could not have been conceptualized. The idea of zoning itself, however, had clearly been derived from the trade codes regulating industrial nuisance.

TURNING A DEAF EAR TO INDUSTRIAL HEARING LOSS

In 1887, the Dutch parliament made an inquiry into the alarming working conditions in the nation's factories that revealed that young boys employed in the boilermaking industry often had to stand inside the boilers holding back rivets as they were hammered in. Willem Ansing, a former smith, socialist, and active trade union president was one of the people interviewed. "Imagine," he exclaimed, "what it does to the brains of these children: they stay in the boilers all day long; it drives you crazy. Although I am not a boilermaker, I became deaf of it myself."

"I noticed!," the president of the committee exclaimed, apparently responding to Ansing's raised voice. Another parliamentary committee member seemed to be unimpressed by the union leader's horror story. "I admit," he said, "that it is not desirable to leave a boy in a boiler for an entire day. Yet I am familiar with men who have done this work from the age of 13 to 16 . . . and who are very vigorous workers now. . . . But wouldn't it be possible to do something about it, for instance by putting wadding into one's ears? . . . Don't you know of bosses who do something in order to fight the adverse consequences?" "No," Ansing answered. And "to put wadding into one's ears won't do, since the boys also need to listen to the things the men standing outside say to them" (Giele 1981 I: 118, 327–328, 365, 366–367).

Another interviewee, Heinrich Struve, a forty-year-old engineer, didn't see the problem at all. For two years he had maintained an office in a boiler workshop with about a hundred workers. "Yet," he stated, "it did not do any harm, I even regretted the moments the knocking was absent." A manufacturer of boilers, Hendrik Suyver, aged fifty-two, and a former boiler-boy himself, needed only a few words to answer the questions of the link between his past professions and his impaired hearing: "All of us get deaf" (Giele 1981 I: 64, 361, 366).

That the engineer did not "hear" the problem of noise is understandable. Although his office was located in the workshop, he did not have to enter the boilers. Neither is it surprising that a manufacturer was relaxed about getting hard of hearing: the noise was a sign of the success of his business. What is remarkable, though, is that even the union leader, whose clear aim was to draw attention to the poor working conditions in Dutch industries, declared the use of wadding to be useless. His observation about the need to stay in touch with the people hammering on the outside of the boiler provides a straightforward explanation for his view, and one that emphasized the chief significance of the act of listening on the shop floor. Similarly revealing is the committee members' claim that boiler boys had grown up into "vigorous" workers, suggesting that the ability to stand the noise was a sign of toughness.

The remarks of both men point to the relevance of the symbolism of noise and cultures of listening for understanding why the use of earplugs was not unambiguously welcomed on the shop floor. Ever since the eighteenth century, physicians had claimed that the noise of hammering and other industrial activities might induce hearing loss. Protecting workers' hearing by the use of earplugs was one of the solutions proposed particularly after the Second World War. Employees, however, often declined medical doctors' advice. The Berlin pharmacist Maximilian Negwer introduced the earplug *Ohropax* as early as in 1907, and made it into a commercial success (Payer 2006: 10). Workers were not among his most enthusiastic customers, though. This section aims to understand the clash between experts' definition of the industrial noise problem, most notably physicians, and the shop floor worker's dramatization of industrial sound. As will become clear, the clash, again, contributed to the persistence of noise—industrial noise in this case—on the public agenda.

That noise could inflict potential damage on the ear was a claim that physicians had already made in the fifteenth century (Dembe 1996: 163). The danger of sudden and traumatic deafness connected with gunpowder was mentioned as early as the late sixteenth century, and coppersmiths and blacksmiths were among the first groups of workers identified as getting hard of hearing because of their professional activities. Among the most well known of such claims are Bernadino Ramazzini's remarks about coppersmiths working in Venice in his *De Morbis Artificum* (Diseases of Workers) published in 1713. The smiths, Ramazzini explained, "are engaged all day in hammering copper to make it ductile. . . . From this quarter there rises such a terrible din, that only these workers have shops and homes there; all others flee from that highly disagreeable locality. . . . [As a result], the ears are injured by that perpetual din, and in fact the whole head, inevitably, so that workers of this class become hard of hearing and, if they grow old at this work, completely deaf" (Ramazzini quoted in Rosen 1974: 514).

From the 1830s onward, studies on "blacksmiths' disease" and subsequently on artillery men, sheet-iron workers, miners, coppersmiths, coopers, millers, metalworkers, locksmiths, boilermakers, and railway workers appeared in European and American medical journals (Dembe 1996; Krömer 1981; Neisius 1989). Over the course of nineteenth and twentieth centuries, new occupational groups, such as telephone-switchboard operators, were added. By 1938, "over 560 occupations" had been classified by the Detroit Health Department as "noisy occupations" (Rosen 1974: 514). And physicians continued to issue warnings about the dangers of noise-induced hearing loss (Güttich 1965; Bell 1966; Burns and Robinson 1970; Saul 1996b: 152–153).

Quieting the machines that made these occupations noisy may seem to be a logical response to the problem of hearing loss. Yet, as medical historian Allard Dembe has suggested for the United States, such noise control may not only have resulted from worries about the danger of noise-induced hearing loss, but also from concerns about efficiency. At the turn of the twentieth century, engineers began to reduce noise from production equipment. "The early attempts to control noise at its source through physical means," Dembe claims, "are unusual compared to other occupational diseases, where greater initial reliance was placed on the wearing of personal protective devices by workers. This may have been due to a realization by engineers and industrialists that noisy machinery often is an indication of mechanical inefficiency that ultimately can result in lower productivity and increase cost" (Dembe 1996: 195).

Wearing "personal protective devices in or around the ears to control exposure to industrial noise," however, did not become common until at least the 1950s. In the United States, this occurred only after a rapid growth in workers' compensation claims, which itself was partly a consequence of the recognition of war-related hearing loss suffered by forty thousand World War II veterans (Dembe 1996: 209). The adoption of legislation that imposed industrial noise standards by setting limit values for noise was delayed right up until the introduction of federal noise standards in 1969. In Europe, financial compensation of occupational hearing loss came even later. Apart from Germany, Sweden, and the Soviet Union, before the mid-1960s, most European countries provided compensation for hearing loss solely in cases where hearing problems affected the ability to work (Dembe 1996: 203; Passchier-Vermeer 1969: 282; Kerse 1975: 121; Neisius 1989). Before the Second World War, the Netherlands introduced legislation stipulating that workers should be provided with some means of ear protection and that their use of earplugs should be assessed. Yet, codes of practice and legislation proclaiming such practices and noise exposure time limits did not become established before the 1960s in Germany, and in the 1970s in the Netherlands and the U.K. (Kuiper 1972: 42–43, 45; Wiethaup 1966; Code 1972; Kerse 1975).

One reason for the delay in legislation regulating hearing protection and compensation, Dembe argues, was the need of evidence about how occupational sounds were linked to hearing loss. Such proof could be gathered only using large-scale audiometric testing, which began in the 1930s—a few years after the introduction of the decibel in the 1920s. Even as late as 1963, a report commissioned by the British government claimed that "the present knowledge of this complex problem," such as the variation in susceptibility of individuals to hearing loss, provided no "sufficient basis for legislation" (Wilson 1963: 128).

In the United States, legislation was further impeded by the fact that "for most occupations, a partial loss of hearing [was] not critical to the performance of the job" (Dembe 1996: 203). Employers' perceptions of the costs of hearing loss prevention and the deep economic crisis of the 1930s made matters worse. In 1929, a German study defined hearing loss as an occupation-related disease in cases where a worker was unable to understand colloquial speech at a distance greater than four meters. Under this condition, a worker should be entitled to receive a moderate allowance as compensation. The authors of the study considered this a substantial economic burden, however, and identified the risk of simulation by workers. As a result, in the revised German Workmen's Compensation Act of 1929, hearing protection was included only in the metals industries (Braun 1998: 255–256; Beck and Holzmann 1929: 31; Neisius 1989: 64–65).

Also unhelpful for large-scale interventions was the fact that the noise abatement societies of the 1910s and the 1920s, which were dominated by the bourgeoisie, gave a higher priority to street rather than industrial noise. For instance, at the end of the 1920s, the Noise Commission of London claimed that street noise was a much more serious problem than industrial noise, because street noise, unlike industrial noise, had no rhythm. This made the clatter of traffic harder to adjust to than the cadence of industrial machines (Brown et al. 1930: 18, 106–107). The people behind such noise abatement initiatives thus legitimized their priorities by comparing street noise's lack of rhythm with the putative regularity of industrial noise. According to Dembe, such symbolism of sound also hampered recognition of noise-induced hearing loss as a disorder warranting compensation in the United States. The positive connotations of loud industrial sounds mentioned in the previous chapter—strength, progress, prosperity, and prowess—complicated the case for obligatory industrial noise control, hearing protection, and financial compensation for hearing loss. And as physicians would discover from the 1960s onward, this not only explained the perspective of industrialists, but also in part the behavior of the workers themselves.

In the meantime, German and Dutch physicians and industrial hygienists focused on educating workers about the need to use earplugs. As early as 1928, the German Society of Industrial Hygiene published a pamphlet on industrial noise and the prevention of hearing loss. The pamphlet explained to workers that "impairment of hearing caused by continuous, moderate occupational noise develops gradually and is first hardly noticed." Initially, the ability to perceive high frequencies would become affected. Later, "the perception of lower tones" might become problematic. Workers should not stop their ears "with plugs of cotton or of flax fiber," but must use "compact plugs made of gauze saturated with petroleum or wax." They should also not postpone consulting a physician "until serious disturbances become noticeable, such as ringing in the ears or attacks of dizziness" (Combat 1929: 2119). The pamphlet also noted that industrial noise on the shop floor should be reduced. Yet the pamphlet did not discuss in detail how this should be done, an approach that was more in line with employers' preference for individual hearing protection than with that of trade unions for more comprehensive noise abatement measures (Neisius 1989:

56, 60). "Relaxation in quiet surroundings when away from work is the most helpful means of preventing ear troubles," the pamphlet concluded (Combat 1929: 2119).

In the Netherlands, physicians taught workers similar kinds of lessons. Through audio-metric testing they revealed the problems of such diverse groups as concrete workers, boilermakers, and weavers. Yet, laborers, an industrial hygienist claimed during a debate on industrial noise in 1958, "show hardly any inclination to use ear plugs." In order to improve this, he tried to "cultivate" a small group of workers to wear earplugs permanently as an example to others, particularly young workers. But it was not an easy task. "We continuously have to harp on the same string," he added, without a trace of irony. An engineer who also contributed to the discussion even claimed that workers, unconsciously or not, created more noise than was strictly necessary. The workers were in dire need of education (Lawaai 1958: 643–646; van Leeuwen 1958).

A few months later, a short movie entitled *Dangerous Noise* premiered in The Hague. The film had been commissioned by the Dutch Department of Social Affairs and Public Health and was aimed at providing workers with knowledge about the causes and consequences of hearing impairment. The problem was, as the head of the Department of Social Medicine at the Netherlands Institute for Preventive Medicine said when introducing the movie to the audience, that being hard of hearing wasn't painful. Staying in noisy environments, in fact, became less annoying to the victims over time: they simply did not hear the noise any more. Even worse, experienced workers ridiculed the use of earplugs. "Elderly weavers jeer at youthful colleagues who wish to protect themselves against the noise. A nail boy wearing earplugs will never become a man" (Bonjer 1959: 733). In response, the film attempted to dramatize the danger of impaired hearing and deafness in everyday life by showing a near fatal accident. It also explained the workings of the ear and the character of audio-metric examinations. In addition, the film mentioned industrial noise control through physical means, yet underlined that when this proved insufficient, one simply had to wear earplugs. Such was the plight of the employees. According to one of the film's medical advisers, it had been hard to plot a course between being scientifically accurate and educational. They believed they had succeeded though.

Although satisfied with the message of *Dangerous Noise,* Dutch doctors kept complaining about the fact that workers refused to wear earplugs. In their view, wearing earplugs was only "a small inconvenience." Yet, of the hundreds of plastic earplugs provided to workers at a weaving factory, only a few were actually used for extended periods. According to the physicians, the workers simply did not understand their personal "interest" in hearing protection. Under such conditions, they grumbled, one could hardly expect employers to invest in noise abatement (Lawaaibestrijding 1959: 740). To the doctors' surprise, workers considered industrial deafness as a fact of life (Lammers 1964: 248).

As late as 1972, Dutch physicians stressed that it was hard to sustain the use of earplugs on the shop floor (Passchier-Vermeer 1972: 26). German publications similarly pointed to workers' opposition to personal hearing protection (Kurtze 1964: 424; Wiethaup 1967: 573). Physicians considered their way of defining the problem of industrial noise crystal clear, and implicitly blamed workers for not speaking their abstract language of early causes and late effects. In contrast, they barely grasped how sound spoke to the workers themselves.

It took years before researchers hit on the idea to actually examine—by talking and listening to the workers themselves—why men and women on the shop floor were less than enthusiastic about wearing earplugs. It was only after this change in strategy that the public voice of workers sang more tunes than mere obstinate objections to hearing protection. Delving into the world of workers suddenly amplified the positive meanings industrial sound could embody for those working on the shop floor.

A study published in 1960 showed that hearing protection made workers feel insecure about the direction from which sounds came, caused communication problems and sometimes "a nasty feeling." The workers also reported that wearing earplugs meant "a kind of embarrassment" in relation to their "comrades." Moreover, they did not believe earplugs to be of any help and considered cleaning them too much trouble. In contrast, the sound of machines represented important opportunities. Industrial noise, for instance, enabled young workers to "sing away to their heart's content" without disturbing their fellow workers. Just as significant was that the noise acted as a kind of reassurance: the sounds signified that the machines were running normally (Invloed 1960: 24–25).

The study reporting these workers' experiences did not go into the details of such reassuring sounds. Other primary and secondary sources, however, suggest what mechanical sounds could say to those operating the machines. As we have seen, engineers might consider industrial noise as a sign of inefficiently running machines. What's more, the specific character of each mechanical noise furnished information about the causes of a machine's inefficiency. The practice of listening to machines in order to diagnose the origins of their mechanical faults was also evident in car repair. At a Dutch antinoise meeting in the mid-1930s, an engineer talked at length about the rattling, puffing, whistling, clicking, tapping, crashing, screeching, howling, crying, grinding, cracking, sneezing, and whizzing of cars. He asserted that the analysis of motor sounds could reveal deviations from normal function before they could be detected visually (Verslag 1936: 77).

Car handbooks for the early generations of mass motorists often included sections that explained to drivers how to listen to their cars. These books literally clarified how the car "spoke" to drivers. A "sneezing" car had something to "tell" about an obstruction in the spray nozzle of the tube of a carburetor's mixing chamber. A "trained ear" could hear that this sneezing never started overnight, yet announced itself by a change in an engine's pitch and rhythm. "Pinking" and "conking" voiced more serious problems. In principle, drivers should mistrust every peep or crack as potential signs of damage (Meyer, n.d.: 19, 20, 22, 24; Ganzevoort 1955: 99–100; Dillenburger 1957: 301–303; de Graaf 1961: 109–113).

Listening to machines was thus not a practice confined to factory life. And while unusual noises suggested mechanical faults, familiar sounds were a comfort to both drivers and laborers: they indicated that the machines were behaving as they were supposed to. The reassuring effects of machines sounding normal are also known from the work of the sociologist of science, Cyrus Mody, and

the historian of science, Gerard Alberts, on the sounds produced in laboratory environments. "When things run smoothly," Mody claims, "these sounds
unfold regularly, marking out the running of a clean experiment. Learning
these sounds, and the experimental rhythm they indicate, is part of learning the
proper use of the instrument" (Mody 2005: 186). Some laboratory employees,
Mody adds, consider data such as those expressing "periodicity" to be much
better processed with one's ears than with the eyes, making the sounds "epis-
temologically relevant" (Mody 2005: 186, 188). Similarly, Gerard Alberts has
explained how in the 1950s, operators of pioneering computers at the Philips
Physics Laboratory in the Netherlands amplified the sounds of their equipment
because they missed the "trustworthy" rattling sounds of mechanical calcula-
tors. Adding amplifiers and loudspeakers to their comparatively silent computers
created an "auditory monitor," thereby restoring an auditory relation with the
equipment (Alberts 2003: 17, 23; Alberts 2000: 9).

Yet, the sounds that machines made not only warned or reassured their
operators. To piece rate workers (workers paid on the basis of their output), the
German physician Gunther Lehmann claimed in 1961, a higher level of noise
simply stood for an increase in income, while lower levels indicated reduced
earnings (Lehmann 1961: 10). The historian Mark Smith has recently stressed
that the same phenomenon explains the ambivalence of workers in the early-
nineteenth-century North American textile industry in relation to the noise of
machines. He mentions a female factory worker who felt intimidated by the
"great groaning joints and wizzing fan" of a new machine, and claims that most
factory operatives aimed to be far from the deafening noise in their free time.
Workers, however, enjoyed the very same noise as an indicator of the employment opportunities that allowed them to share in the region's progress (Smith
2001: 136, 137, 140–141).

Moreover, and in contrast to Smith's female worker, men often appreciated noise for its loudness. For whoever controlled the noisiest machine on the
shop floor, Gunther Lehmann argued, noise was the expression of his importance and his power over others. This was similar to the meaning noise possessed
for young drivers, he added, as an expression of a motorist's strength and mas-

culinity (Lehmann 1961: 10). This may explain the embarrassment workers felt when wearing earplugs: they did no good at all for one's place in the masculine hierarchy of the shop floor. How significant such signs of toughness could be is also illustrated by a study of British shop floor cultures in the 1970s. This study quoted a foundryman who proudly claimed that his work "on the big hammer," a "six-tonner" was "bloody noisy," but explained that he had gotten used to it. At home he was considered to be a "silly deaf old codger." He could "hear perfectly well in the factory," though. "If I see two managers at the end of the shop, I know . . . just about what they're saying to each other" (Willis 1980: 189–190).

According to a Dutch industrial hygienist, employees were much more afraid of losing their ability to see than their ability to hear, since hearing loss only became a social handicap after many years. Furthermore, wearing earplugs could be painful or irritating to the ear, and using them was an attention-taking and time-consuming affair. Workers needed to take care in cleaning them, to store them correctly, and not to lose them. Since laborers such as weavers were piece rate laborers, taking care of earplugs was simply "too much." Moreover, being remiss in using earplugs had no consequences for a group of weavers as a whole, whereas leaving weaving looms untidy did. The industrial hygienist therefore experimented with hearing protection that the weavers could easily obtain from an automaton and throw away afterward. The automaton also reduced their sense of being controlled by their bosses. After its introduction, the acceptance rate rose to about 70 percent (Lammers 1964: 249–251).

Workers supposed, as industrial hygienists regretted in the early 1970s, that earplugs reduced their capacity to understand speech. The hygienists claimed that theoretically this was impossible since the signal-to-noise ratio remained the same with or without plugs. To their surprise, however, a subsequent study of this phenomenon substantiated the workers' knowledge. This was particularly true for those workers already suffering from hearing loss, which the authors put down to problems with recognizing high-frequency consonants in signals below a certain level. They concluded that workers should be taught to shout even if they did not hear much noise because of their earplugs (Lindeman and van Leeuwen 1973). Taking the workers' experiences seriously thus revealed the workers "indigenous knowledge"—in the definition of local knowledge—of sound perception on the shop floor (Watson-Verran and Turnbull 1995).

In sum, until the legislation establishing noise exposure limits in the late 1960s and 1970s, it was left largely up to the industrial workers themselves whether or not to use hearing protection. Physicians, however, did not succeed in defining and dramatizing the problem of industrial hearing loss in a way that convinced workers to take precautionary measures. Their notions of today's risks and future problems clashed with the workers' cultural values, the meaning sound had for them, and their routines on the shop floor. Medical experts combined an abstract discourse about hearing impairment with an individualizing approach, which considered the workers responsible for their own ears. The workers, however, turned a deaf ear to the physicians' rhetoric. To them, the hum of a machine could be a comforting sound since it informed them of its proper functioning. The noise enabled them to sing as loud as they wanted and reminded them of their earnings. Putting up with the noise enabled them to parade their toughness, and controlling a noisy machine signified a position of power. They wanted to communicate with their comrades, not to indulge in time-consuming cleaning and storing of earplugs when the collectivity of workers demanded something else. Cultural meanings of sound—both expressed in a particular symbolism of noise and a shop floor culture of listening to machines—thus largely explain workers' lack of enthusiasm for hearing protection. On occasions where workers dramatized industrial sound, they referred to its uplifting meanings, and invoked the auditory topoi of the sensational and the comforting sound.

Once again, the dramatization of an industrial noise problem did not match the language of the world in which it aimed to intervene. In the case of the nuisance hearings, the dramatizations of complainants failed to connect with the problem definition of the experts: the makers and executors of the nuisance legislation. In the example of industrial hearing loss, its definition and dramatization by medical experts failed to touch workers. This is not the only explanation for the persistence of the public problem of industrial hearing loss.

Physicians followed an individualizing strategy, employers balanced the costs of interventions against the costs of doing nothing—such as paying for disability or a reduced mechanical efficiency—and governments legitimized their postponement of legislation by referring to scientific uncertainties. Yet, the clash between physicians' definition and dramatization of industrial hearing loss and the meanings of industrial sounds to the workers was also highly significant to the persistence of noise as a pubic problem.

INDUSTRIAL NOISE CONTROL AND MASKING NOISE BY MUSIC

As we have seen, the first pleas for reducing the noise of industrial machines that appeared in the years around 1900 did not only follow from anxiety about occupational hearing loss, but also from concerns about mechanical efficiency. Enhancing the efficiency of machines was not the sole incentive for industrial noise control, however. Enhancing the efficiency of workers was another.

As early as 1913, Josephine Goldmark, chair of the Committee on the Legal Defense of Labor Laws of the U.S. National Consumers' League, was quite clear about the dangers of noise. Noise, she claimed, "not only distracts attention but necessitates a greater exertion of intensity or conscious application, thereby hastening the onset of fatigue of the attention." According to her description of the textile trade, some kinds of motor sewing machines "set almost 4000 stitches a minute. Let any observer enter a modern roaring, vibrating workroom where several hundred young women are gathered together, each at her marvelous machine, which automatically hems, tucks, cords, sews seams together.... Her attention cannot relax a second while the machine runs its deafening course.... The roar of the machines is so great that one can hardly make oneself heard by shouting to the person who stands beside one" (Goldmark 1913: 54).

In the context of the Taylorization of western industrial life in the 1920s and 1930s, suggestions of inefficiency in workers operating noisy machines clearly rang a bell. One type of response was the confinement of the shop floor noise. Sound deadening foundations that reduced the vibration of machines and the exclusion of machinery noise from particular areas using insulation were discussed and introduced (Reducing 1921; Richardson 1927; Davis 1934; Wagner 1936a; Thompson 2002). Quieting machines by suppressing the causes of noise at their source was usually seen as the most effective approach though. This focused on oiling moving parts, fixing defective ones, improving gearing, balancing machinery, reducing vibration and resonance caused by magnetic pulsation, and preventing sudden discontinuities in the motion of machines (Abbott 1936; Wigge 1936; Ballot 1948; Zeller 1950).

The other type of response was to enhance employees' efficiency by restoring the sensation of rhythm on the shop floor. While the Noise Commission of London stressed that street noise was chaotic and industrial noise rhythmical, experts focusing on industrial noise took quite a different stand. Although work had once been rhythmical, the introduction of modern machines had brusquely unbalanced the sound coming from the shop floor. The beat of the machines was more rapid and fixed than human rhythms. Even worse, a standard workshop had not one, but a multitude of machines, with varying rhythms that, taken together, produced sonic chaos. Clearly, the experts' experience of factory sounds departed from the anti–street noise activists' ideas of industrial noise. Yet the arguments underpinning their positions were remarkably similar: rhythmic sound was the thing to strive for, whereas the absence of unambiguous rhythm—the most dangerous form of noise—was the situation that needed to be avoided. Playing the phonograph, the gramophone, or the radio on the shop floor was therefore seen as a promising way of recreating a rhythmic sensation within the factory walls.

This strategy grew out of the assumption that the rhythm of music facilitated effective performance. Initially, arguments for the significance of musical rhythm for work were historical-anthropological in character. In 1896, the German historical and anthropological economist Karl Bücher claimed that the origins of labor were to be found in the rhythmical character of physical labor. Since *Naturvölker* (primitive peoples) had a natural disinclination for work, but loved to dance, only rhythm could explain why people had learned to endure simple, fatiguing tasks to an extent beyond that needed for basic survival. From and intertwined with the lust- and fantasy-evoking rhythm of bodily movement, music had evolved to facilitate the performance of individual labor and the synchronization of communal work. Rhythm had thus raised the productivity of labor. Art and technology, now differentiated domains of life, had once been two sides of the same coin (Bücher 1896: 80 ff). As Goldmark put it, "Not the poetry of existence only, but all the daily offices of life—spinning, weaving, sowing the grain, harvesting, and the rest—inspired song and dance, their own rhythms. Even today innumerable survivals persist. . . . In the midst of discordant city traffic, workmen who are mending the pavements drive steel wedges with rhythmic shouts and rhythmic alternating blows of their sledges. They know, instinctively, that the rhythm makes the work easier" (Goldmark 1913: 80).

In contrast, the fast movements of modern machines, Bücher argued, produced a "confusing, deafening noise" in which one could "hear" but not "experience" rhythm, thus evoking "merely a sense of frustration" (Bücher 1896: 115). And, as Goldmark added, "Not only is the beat of the machine much more rapid and regular than the more elastic human rhythms; it is often wholly lost in the chaos of different rhythms of the various machines, belts, and pulleys in one workroom" (Goldmark 1913: 82). In the future, Bücher suggested, one might hope to succeed in combining technology and art once more "into a higher rhythmical unity" (Bücher 1896: 116).

It is not immediately obvious what unity of technology and art Bücher had in mind. Yet, in a sense, Thomas Edison appeared to have made his dreams come true. In 1915, the historian of music Joseph Lanza claims, Edison "used a programmed selection of phonographic music for factories to determine the extent to which it would mask hazardous drones and boost morale. But the infant loudspeaker and transmission technology was still too weak" (Lanza 1994: 13). The idea itself, however, continued to fascinate industrial psychologists. In his doctoral thesis, *The Influence of Music on Behavior,* published in 1926, the Princeton graduate student Charles M. Diserens, summarized practices that already existed, such as speeding up typewriting with the help of music, as well as recent experiments studying the effects of tones and tonalities on blood circulation and muscular force. The experiments' results were far from straightforward. Yet many reports, often based on questionnaires, claimed simply that music in the workplace led to an increased rate of work, fewer errors, better temperament, and less fatigue. Theoretically, Diserens claimed that these phenomena could be explained by arguing that rhythm "eliminates the strain of voluntary attention" and reduces fatigue "by lending regularity to muscular reaction. . . . It is a conservative factor. Tone, on the other hand, lends force to muscular movements" (Diserens 1926: 122).

All of the numerous reports and studies undertaken in both the United States and Europe that had succeeded those discussed by Diserens claimed an increase in output under the influence in music (Wyatt and Langdon 1937; Antrim 1943; Burris-Meyer 1943; Prause 1948; Muziek 1955). "It diverted the mind from the monotonous conditions of work, provided an attractive aim, made time seem to pass more quickly, and created a more cheerful attitude towards work," one famous British study from 1937 explained (Wyatt and Langdon 1937: 40). It took some discussion, however, to decide what *kind* of music was most suited to the job. The quick rhythm of one-steps coincided with high output, yet made "the workers feel 'lively and restless,'" and the rhythm of marches "clashed with some of the movements involved" in factory work (Wyatt and Langdon 1937: 40, 42).

According to a report Philips published in 1938, the music used should not be too sentimental since the daydreams of the factory girls were already leaning too much toward the sentimental and sexual. On the other hand, jazz was not appropriate for the shop floor because of its "sudden variations in high and low tones" and the "very pronounced rhythm" that conflicted with the rhythm of work and led to the "stammering" of the hands and feet (Opdenberg [1938]: 6).[6] Similarly, the people behind the BBC program *Music While You Work* banned "hot" music with complex rhythms because it was thought that it would create "a confusion of sound" (Reynolds 1942: 6–7; Korczynski and Jones 2006). (Figure 3.1.)

In 1950, a Dutch industrial hygienist stressed that precisely because of the rhythmical character of factory work, the music used must be "more than rhythm." "A clear, easily appealing melody should expound, as it were, the

Figure 3.1 A music installation on a shop
floor, 1930s. The music installation comprises
a radio receiver, a gramophone, and an
amplifier. A microphone serves for transmitting
announcements. From H. Opdenberg, *Music
in Worktime* ([Eindhoven]: Philips, [1938]), p. 9.
Courtesy Philips Concern Archives.

difference from the noise." By all means music needed "to create order in the chaos of sound" (Schröder 1950: 191–192). Yet, as many authors claimed, one did not need to adjust the rhythm of the music to the rhythm of the machines. Nor would employing fast music necessarily speed up work. Music, it was widely believed, influenced workers more by claiming their attention and affecting mood than through its tempo. The speed of work simply varied too much to be dependent on the tempo of the music (Groen 1948; Reynolds 1942; Halpin 1943). Moreover, both European and American studies had found that the clatter of machines seemed to adjust to the rhythm of the music, and that even amid terrific din, the music stood out (Bergius 1939). As the contemporary pianist and music journalist Doron K. Antrim clarified, "the ear tends to follow" the agreeable "regular tonal pulsations" of music and to "forget" the irritating and fatiguing "irregular pulsations" of noise (Antrim 1943: 276).

Until the introduction of the compression amplifier in the second half of the 1940s, the power of amplifiers was often insufficient to mask the noise of machines.[7] The adherents of shop floor music did not seem to be bothered too much by such technical problems though. Even if the "hum" and "shrieks" of machines mixed with the sound of music, a Dutch expert in industrial efficiency asserted, the music still had a positive effect (van der Toorn 1949: 206). Music of relatively high frequency, others explained, did not have to be very loud in order to be audible above the low-frequency noise produced by many machines (Schröder 1950: 191; Cardinell 1944).

Many factory managers enthusiastically embraced music in work time. Interest in high production levels during the war further stimulated the rise of factory music (Cardinell 1943; Hajduk 2003). For instance, the introduction of playing the radio program *Music While You Work* on the shop floor in 1940 started from the idea of maintaining British morale in the terrifying early days of the war (Reynolds 1942). And the Muzak Corporation, created in 1934, aimed to boost morale both during and after the war (Selvin 1943; Barnes 1988; Husch 1984; Lanza 1994). In the late 1940s, it launched Stimulus Progression, the practice of sequencing compositions in such a way that the music effectively countered employees' waves of fatigue (MacLeod 1979). Today, the Muzak Corporation claims that one hundred million people hear its music worldwide, from factories and offices to shops and malls.[8]

It was not that no one questioned the putative blessings of music on the shop floor. Some Swiss and German authors considered it irresponsible to mask noise by music without abating the noise itself or providing hearing protection for workers (Brunner 1952; Blokland 1958b). Yet, as an American critic was forced to admit, the music seemed to be "generally accepted as a desirable thing" (Pepinsky 1944: 179). Judging by the published questionnaires and the success of radio programs such as *Music While You Work*, this was also true for workers (Blokland 1958a; de Groot 1981). Workers even began introducing audio sets on the shop floor themselves (Lanza 1994: 13; Cardinell 1944). In the Netherlands, Philips employees explicitly asked for *Labor Vitamins*, a program modeled on *Music While You Work*, and demanded popular instead of classical music (Bijsterveld 2002).

Managers tried to restore the rhythm of work within the chaos of mechanical din, thus following a long-standing symbolism of sound that associated noise with chaos and rhythm with order. They also contributed to a culture of listening in which listening to the radio became increasingly combined with other activities—something already quite common, as Susan Douglas has claimed also occurred in North America by the 1930s (Douglas 1999: 84). It is, in fact, one of the ironies of history that background music, introduced as a means of masking noise, has become one of the practices that noise abatement societies such as Pipe Down International today deeply yearn to get rid of.

CONCLUSIONS

The history of music on the shop floor is the only episode in the history of industrial noise discussed above in which the dramatization of the noise problem formulated by experts matched what laymen considered their everyday troubles. Social scientists emphasized that the chaotic rhythm of mechanical noise caused fatigue among workers. Enhancing employees' efficiency by restoring the sensation of rhythm on the shop floor was their solution—a remedy welcomed by workers as an antidote to boredom. In the two other examples, the definitions and dramatizations of industrial noise produced by experts clashed deeply with the ways in which non-experts phrased their experiences of sound and noise. In the nuisance law hearings, laymen saw their attempts to stand up against noise often frustrated by the local authorities' interpretations of legislation. On the shop floor, physicians' efforts to convince workers of the risks of industrial hearing did not succeed. In both cases, the clashes contributed to the persistence of noise as a public problem. In the story of music in industry, all parties shared the idea that rhythm could remedy noise, yet the end result—the diffusion of background music—opened up a new noise problem for others.

In the trade code case, complainants employed the topos of the intrusive sound, first identified in chapter 2, to dramatize their experiences—dramatizations that did not fit into the codes' definition of noise, which in practice boiled down to the damage noise could produce, particularly damage to property. In the discussion about earplugs, workers invoked the topos of the comforting sound to reject the solutions implied by the physicians' definition of the industrial hearing loss problem. Intriguingly, it was the rhythm of machines that reassured them, just as it was the rhythm of music in industry that workers enjoyed hearing, because of its apparent ability to cut through the sound of the machines.

If we bring back into mind the most common explanations for the persistence of noise on the public agenda, as discussed in chapter 1, it becomes clear that the persistence of the industrial noise problem was not due solely to economic growth at the expense of health interests. Not every new machine was noisier than its forerunner, and many of the noisy hammering trades had disappeared over time. The urge for industrial growth certainly fostered the issue of trade licenses. Yet nuisance laws themselves were the result of property owners' attempts to recover "the price of noise" to industrialists, whereas industrialists aimed to increase the efficiency of machines through noise control and the efficiency of workers by masking noise. And vice versa, noise abatement was not merely inspired by health interests. Health was a relevant argument in defending the ears of workers and the peace of hospitals, yet for churches and schools, cultural traditions that associated learning and contemplation with silence and solitude helped these institutions to be established as islands of silence. It is even more difficult to see how the visual character of western society might have contributed to the persistence of the industrial noise problem: listening cultures, in fact, happened to be crucially significant in cases of industrial hearing loss and factory music.

As I have emphasized, the most significant legacy of the trade codes was the idea of establishing separate zones of industrial activity and zones of tranquility. These spatial solutions, the basis for many noise abatement policies to come, accommodated conservative interests in land and tenement protection, liberal aspirations to expand industry, and the cultural heritage of keeping the places of learning, religion, and recovery relatively quiet. The regionalization of noise pollution, however, created the need to commute between residential and industrial areas. The roar of traffic, indeed, was to become the next topic in the continuing debate about noise.

INSTRUMENTS OF TORTURE:
TRAFFIC NOISE AS UNCIVILIZED BEHAVIOR

THE CITY OF DIN

The motor-horn! The motor-horn! I often wonder why in all the world such an instrument of torture has ever been permitted to exist even for a single day! But there it is ... and of a variety...! First ... comes the hoarse reedy *squawk* that betokens the cheaper car, the taxi, and the like, croaking like some gigantic raven from a Dinosaurian age as the driver dashes round a corner or threatens a slow-going horse-vehicle in front of him.... Then follow horns of a more ambitious ... quality; some of them passing for "musical" ... like the "Gabriel" Horn, whose sounding diapason is startling enough, in all conscience, to awaken the dead.... And at the end of the list come unearthly screeches, squeaks, and groans, from the various noise producers "on the exhaust." (McKenzie 1916: 33–35)

This detailed description of the sounds of the motor-horn is quoted from *The City of Din: A Tirade against Noise.* The author, Dan McKenzie, a British surgeon, published the book in 1916, in the midst of the Great War. He included a relative's experience of the "unceasing, heartrending, brain-rending" noise of this war (McKenzie 1916: 102). Yet, the book's main topic was urban din. His tirade was only one out of hundreds of pamphlets, essays, reports, journal articles, and newspaper items to appear in print in Europe from the late nineteenth century onward, claiming that the "nerve-racking" noise of modern life, and of contemporary city life in particular, had become unbearable. Amid the myriad of published complaints, concerns about street noise, or the noise of traffic—trains, trams, automobiles, motorcycles, trucks, buses—predominated. Among the protests against traffic noise, those against the car and the motorhorn were by far the loudest. In the wake of these outcries over urban noise, antinoise societies were created, giving voice to their constituents' complaints through campaigns, conferences and exhibitions.

The antinoise actions before World War II illustrate how the sounds that accompanied new technologies were considered to cut through the existing societal order. While industrial noise emission had largely been handled as threatening property and institutions, early-twentieth-century street noise was widely understood to endanger the minds and bodies of city dwellers. Initially, the struggle for silence was phrased mainly in terms of the "civilized" needs of the intellectual elite against the "indifferent," "irrational," and "barbarous," behavior of the masses, whereas noise-abaters showed little hostility toward technology itself. Even in the later phases of the noise abatement movement, when city noise, expressed in the language of scales of loudness, increasingly came to be defined as a general health hazard that negatively affected the efficiency and productive capacity of all city dwellers, making noise continued to be associated with a lack of manners and keeping silence with proper behavior.

Consequently, public education came to be seen as the alpha and omega of tackling the city noise problem. Although changes in law, technology, and space, such as traffic regulation, alternative pavements, new transportation constructions, and city planning were also proposed and executed, public education continued to be seen as the ultimate way of creating silence. This enabled an intriguing discourse coalition to be made between noise abater societies and those of automobile owners, which facilitated the creation of new forms of order, integration, and enhanced predictability—a new smooth and controlled rhythm, so to speak—in city life. Tragically, the result was an endless flow of traffic. It was only after the process of canalizing traffic that the employment of new units and techniques of measurement led to a significant conceptual change. The display of the never-ending stream of cars in terms of decibels entailed a reconceptualization of traffic noise from a *chaos* of unwelcome sounds produced by badly behaving *individuals* to *levels* of unwanted sound resulting from a *collective* source. This, in turn, would open the gateway to new kinds of noise abatement after the war.

THE CITY NOISE PROBLEM: CIVILIZATION VERSUS BARBARISM

Between the second half of the nineteenth century and at the beginning of the twentieth, many of the complaints about city noise that were published, at least in Europe, came from intellectuals. They generally considered noise to be a brute assault on their mental refinement. A lament by the philosopher Arthur Schopenhauer, published in 1851, is the most famous charge against noise—an ever recurring *Leitmotiv* in the noise abatement literature. Schopenhauer regarded the cracking of whips as a hideous sound, distracting him from his philosophical work. Whip-cracking, he complained, made "a peaceful life impossible; it puts an end to all quiet thought.... No one with anything like an idea in his head can avoid a feeling of actual pain at this sudden, sharp crack, which paralyzes the brain, rends the thread of reflection, and murders thought" (Schopenhauer 1974/1851: 157).[1] According to Schopenhauer, many people showed no sensitivity to noise. However, such people were likewise insensible to "arguments, ideas, poetry and art—in sum, to mental impressions of all kinds, due to the tough and rude texture of their brains" (Schopenhauer 1974/1851: 156).

The British mathematician Charles Babbage, who campaigned against the nuisance of London street music, phrased his complaints in a similar way. He rejoiced the passing of a bill in 1864 whose aim was to control such noise. "The great encouragers of street music," Babbage wrote, "belong chiefly to the lower classes of society," such as tavern-keepers, servants, visitors from the country and "Ladies of doubtful virtue." The instruments of "torture" that they seemed to love, destroyed "the time and the energies of all the intellectual classes of society" (Babbage 1989/1864: 254, 253). Although he was aware that the invalid and the musical man were also annoyed by such people, Babbage focused his complaints mainly on "the tyranny of the lowest mob" on "intellectual workers" (Babbage 1989/1864: 259). The Austrian ethnologist Michael Haberlandt simply claimed that the more noise a culture could bear, the more "barbarian" it was. In contrast, tranquility was "the womb of all higher intellectuality" (Haberlandt 1900: 177–178).

As has been discussed in the previous chapter, concerns about street noise—including the sounds of the trades—and its effects on men of letters well predated the nineteenth century. Historian Emily Cockayne (2007) has recently documented how urban dwellers in seventeenth- and eighteenth-century England had to cope with a wide variety of street noises—the sounds made by coaches and carts on cobblestones, pigs and dogs, taverns, inns, and alehouses, horns and bells, mills and smithies, and street hawkers and street musicians. Cockayne claims that attitudes shifted toward a heightened sensitivity to noise at the end of the seventeenth century, owing to a swelling population and increased trade. Yet, the noises that were restricted, for instance through licensing, were those made by the poorest classes: the popular entertainers and low-profit traders whose sounds did not fit into the lifestyle of a growing professional urban class longing for peaceful reading and studying.

Other historians and literary scholars situate this change in attitude to street noise in the nineteenth century. Some connect it to the rise of the idea that the bourgeois interior had to be a safe haven from a harsh world (Cowan 2006). Others emphasize that the campaign against street music in London had strong undertones of contempt toward persons from the lower classes and of a foreign origin. These historians see the campaign against street noise, in which Charles Dickens, the historian Thomas Carlyle, and the illustrator Richard Doyle joined forces with Babbage, as the attempt by a rising elite of "authors, artists, actors, and academics" to secure professional standing and the right to work at home without being distracted. The London bill of 1864 could, at least on paper, restrict street music "on account of the interruption of the ordinary occupations or pursuits," whereas a previous statute had allowed only for restrictions on account of illness. In practice, the bill helped the pursuits of the bourgeois, since successful enforcements happened exclusively in middle-class residential areas (Picker 2003: 53; Assael 2003: 194). Cockayne acknowledges this nineteenth-century trend but adds that there is "evidence of the beginnings of this drive from the mid-seventeenth century" (Cockayne 2007: 128).

Schopenhauer, Babbage, and Haberlandt did not mention the roar of motorized traffic, although, as discussed in chapter 2, their contemporaries had already begun lamenting the disturbing sounds of trains. Subsequent intellectuals, however, increasingly transferred the elitist convictions of their predecessors from whip cracks, vocal cries, and street music to the mechanical sounds of the city streets. They spoke, as Peter Bailey (1996) has shown, about a struggle for silence in terms of civilization versus barbarism.

For instance, the philosopher of mind James Sully published an essay entitled *Civilisation and Noise* in 1878. He argued that if "a man wanted to illustrate the glorious gains of civilisation, he could hardly do better, perhaps, than contrast the rude and monotonous sounds which serve the savage as music and the rich and complex world of tones which invite the ear of a cultivated European to ever new and prolonged enjoyment. . . . Yet flattering as this contrast may be to our cultivated vanity, it has another side which is by no means fitted to feed our self-complacency. If the savage is incapable of experiencing the varied and refined delight which is known to our more highly developed ear, he is on the other hand secure from the many torments to which our delicate organs are exposed" (Sully 1878: 704). He described one such torment as the "piercing noise of a train, when brought to a standstill by a break." Another was the "diabolical hooter" used to remind railway workers of "their hour of work." The loudness and harshness of these and comparable sounds of traffic and factories were a student's "proverbial plague." Concentration, the "counteractive force" of civilized man, was not enough to neutralize the effects of the increased sensibility and the irritating impressions amid the dense, and often indifferent, population of the cities (Sully 1878: 706, 720, 707, 709).

———

In 1908, the German professor in philosophy and pedagogics Theodor Lessing was even more outspoken in his essay "Noise: A Lampoon against the Din of Our Lives" (*Der Lärm. Eine Kampfschrift gegen die Geräusche unseres Lebens*). Lessing declared that he was annoyed by both traditional noise, such as the din of church bells and carpet-beating, and the more recent nuisance of rattling machines, shrilling gramophones, ringing telephones, and roaring automobiles, buses, trams, and trains. The latter type of noise, however, was "incomparably worse" than the former and made present-day life "nerve-racking" (Lessing 1908a: 45; Lessing 1908b; Lessing 1909; Wendt 1910). For Lessing as for Sully, noise was profoundly anti-intellectual. Noise, he asserted, raised and exaggerated deeply rooted human instincts and emotions—the "subjective" functions of man's soul—and narrowed and dimmed the intellectual and rational—"objective"—functions of the soul. Noise was the most primitive and the most widely used means of deafening the consciousness. In fact, noise was the "vengeance" of the laborer working with his hands against the brainworker who laid down the law to him. Silence, on the other hand, was the sign of wisdom and justice. "Culture," Lessing stressed, "embodied the genesis of keeping silent" (Lessing 1908a: 11, 20). Or as the writer and journalist Emmy von Dincklage had phrased it three decades earlier: it was "infinitely distinguished" to prevent speaking in a loud voice, whatever the situation (von Dincklage 1879: 414).

In *The City of Din*, Dan McKenzie claimed roughly the same. If his crusade proved successful, he wrote, one of the consequences would be that "the raucous tones of the raucous-minded would give place to the gently-voiced opinions of the mild and tolerant. So that this particular crusade is only one small part of a grand effort at the refinement of the human spirit" (McKenzie 1916: v). "In the long run," the English accounting expert Stanley Rowland proclaimed in 1923, the difference between the noisy and the not-noisy was that "between self-possesion and self-assertion—or, more generally, self-diffusion." By self-diffusion he meant something like a violent broadcasting of a person's sound. One could encounter such self-diffusion at the theater, where people kept "a running commentary on the action," in the "abominable shrieks" of newsboys distributing newspapers, as well as in the "brutal objurgativeness" of the motor-horn, the "outrageous noisiness of the motor cycle," and the "semi-barbarious emotional music" of jazz (Rowland 1923: 315, 317, 318, 319, 314).

In short, pamphlets and essays like those of Sully, Lessing, McKenzie, and Rowland displayed a deep concern with the disrupting effects of noise on societies' intellectual strength and cultural maturity. The higher classes, the refined mind, and a cultivated self-control were widely understood by the social elite to be threatened by both the traditional and the new sounds of the lower classes, vulgar emotions, and brutal self-diffusion. To one of Lessing's' colleagues it was therefore self-evident that the homes of brainworkers required a higher level of protection from noise than those of manual and farm workers (Nussbaum 1912–1913). Remarkably, these intellectuals failed to recognize that some of the new technologies whose noise they abhorred, such as the early automobile, were the privileged property of wealthy gentlemen rather than of the masses (Fraunholz 2003). Their lack of awareness only underlines how the intellectual elite transposed earlier repertoires of dramatizing noise to new urban experiences. They clearly wrapped their discomfort with the new age in a set of long-standing connotations in which silence stood for everything higher in the social hierarchy (unless it was imposed), and noise for everything down the ladder. Moreover, the sounds of individually owned cars and gramophones were less impersonal than those of trains and factories, enabling more direct attacks on the behavior of identifiable groups of people.

Even the machine itself came to be invoked as a metaphor of the mind that stressed the intellectual's refined and vulnerable character. According to Rowland, "the avoidance of friction of the human senses is as important for the equable functioning of the mind as is the elimination of mechanical friction for the proper working of a dynamic machine" (Rowland 1923: 313). For McKenzie, the modern mind was "a delicate instrument, the needle-indicator of which trembles and oscillates to the finest currents of thought and feeling. By culture and education we have acquired the sensibility of the artist or poet. And yet we continue to expose this poised and fragile instrument to the buffeting of a steam-hammer, to the shriek of a locomotive!" (McKenzie 1916: 52). Such claims did not stand alone. Seeing the modern mind as a fine-tuned needle-indicator reflected a more general concern among the social elite about an increase of nervousness and irritability as a consequence of the numerous new urban sensory impressions they encountered (Beuttenmüller 1908; Birkefeld and Jung 1994; Saul 1996a).

This situation was agreed to be most acute in the cities. Between 1900 and 1914, the motorization of cities such as Hannover, where Lessing lived, was still moderate. In 1907, Hannover had 1,472 motorcars—Germany as a whole possessed over twenty-seven thousand motorcars versus about two million horses used for transport (Birkefeld and Jung 1994: 45). Still, the density of traffic rapidly increased. Along a busy location in Hannover in 1900, trams passed 850 times every twenty-hour day, creating the sounds of brakes and metallic creaks of wheels in the curves of rails every ninety seconds (Birkefeld and Jung 1994: 39). Vienna had twelve thousand buildings in 1880 and forty-one thousand in 1910. The multistock buildings intensely reflected the many sounds of horse, steam and electric trams, bicycle bells, and automobiles (Payer 2004: 106). Mass-produced automobiles in the 1910s with all-steel bodies were "fragmented, loosely joined assemblages" and thus "clattering contraptions" that "drummed" on the streets, even apart from their "rough-running engines" (Gartman 1994: 49; Jung 1902).[2] Older types of transport, such as many horse-drawn vehicles, also filled the streets. A 1861 handbook of local ordinances in Amsterdam mentioned that the drivers of these vehicles had to yell before passing another (Calisch 1861: 105, 130). The legacy of such regulations may have contributed to the multitude of auditory signaling there.

McKenzie claimed that in London the quality of the sound on the streets had improved by 1916. They had become far "less clattering, less jarring, less varied" as a result of using smooth-surfaced roads made of asphalt or wood and pitch and rubber tires instead of iron-girded wheels, and electric instead of steam engines. But in quantitative terms, the noise had increased (McKenzie 1916: 32, 105): "The roar of the traffic of motor-buses, taxi-cabs, and motor-cars is of a deeper, more thunderous, and more overpowering nature than in former days, principally because vehicles are heavier and are driven at a much greater speed" (McKenzie 1916: 33). The irregular and unexpected sound of the motor horn, the sounds of gear levers of buses, and the sounds of trains such as "the clank of the wheel at the end of each length of rail," all contributed to this din (McKenzie 1916: 69).

The social and intellectual elite who complained so bitterly about city noise, however, did not consider fleeing the city for silence in a rural life as the ultimate solution. In contrast to the contemporary urban environment, they associated nature with pleasant sound and the urban past with tranquility (Haberlandt 1900; McKenzie 1916). Yet, at the same time, they viewed noise as ubiquitous, which meant that unpleasant sounds could be found even in the countryside. Lessing was one intellectual who hoped to find tranquility in villages. In practice, however, he soon felt haunted by the rural sounds of the carousel, the steam-plough, threshing-floors, boilermakers, and all kinds of animals. Even in the most far-away valley of the Alps, he lamented, one would encounter a gramophone (Lessing 1908a).

Technology itself, however, was not seen as the bad genie. Of course, it had been "left to scientific civilisation to fill the world with stridency," as McKenzie claimed (1916: 28). But not one of the intellectuals advocated that society refrain from technological progress. I therefore do not follow historian Jon Agar's claim that McKenzie and the British Anti-Noise League that was to pick up McKenzie's work endorsed a worldview that was both antimodern and antitechnology (Agar 2002). Although McKenzie heralded the tranquility of nature, was nostalgic about the sonic past, lamented the roar of motorized traffic in the city, and thus colored his argument in antimodern tones, he, like his fellow writers of early antinoise essays, did not oppose technology. This was the case for Lessing as well. While a socialist politically and a conservative culturally, particularly with respect to modernization, he had nothing against technology per se. Lessing considered automobiles, motorcycles, and airplanes as the transportation of the future. For him, it was society itself that had to be reorganized, a view that many contemporary German intellectuals upheld. They believed that the condition of heightened nervousness was a sign not only of societal crisis, but also of necessary societal adaptation (Stoffers 1997).

Instead, noise-abating intellectuals viewed public education as the most fundamental solution to the noise problem, a view that was in keeping with the notion that making noise was socially disruptive and an indication that both the social order and mental life were degenerating. Although such intellectuals considered all kinds of practical measures to be helpful, they stressed that teaching the public how to behave was the best strategy for obtaining lasting results. Haberlandt asserted that in every school an 11th commandment should be taught: "Thou shalt not make noise" (Haberlandt 1900: 182). Rowland argued that the "only hope of ultimate reform would seem to lie with our schools, where the subject of general social deportment might well occupy a more important place" (Rowland 1923: 316). McKenzie believed that much could be done, "But the victory will come all the sooner if public opinion can be educated" (McKenzie 1916: 111).

Such educational measures could be supported by law. Sully pointed out that since the ear, unlike the eye, had no natural defense, the law should recognize and protect the right to silence in one's home. This could be achieved by transforming certain noises into penal offenses. McKenzie wanted to prohibit the use of the automobile horn at night. As Lessing made clear for Germany, however, the law remained inadequate for the abatement of most noises. Since noise could be punished only in cases where it was "generally disturbing," "unnecessary," and "intentional," decisions about exactly what to punish were quite arbitrary. Moreover, the noise of machines could hardly be implicated, since the noise of trains, trams, and factories was supposed to arise from the "nature" of things, and "harm" mattered only in case of damage to "property," thus not to one's strength, health, or ability to sleep (Lessing 1908a: 73–91).

Other intellectuals proposed solutions focused on a spatial reorganization and visualization of city life that often embraced new technologies. The German social psychologist and nerve specialist Willy Hellpach believed that the railway station of his day—1902—was far less noisy than it had been before, a change he attributed to "visualization." Many "intoxicating" aural signals—horns, whistles, shouts, and bells—announcing and accompanying the arrival of trains had been replaced by "sober and non-obtrusive" inscriptions (Hellpach 1902: 32). He welcomed the increasing separation of the home and the workplace, since the noises of manual workers such as the locksmith, cabinet-maker, and cobbler were more disturbing to neighbors than the dimmed, continuous noise of a factory. The real causes of noise, he asserted, could be found in the things that had remained as they were: the narrow, dark streets through which the traffic squirmed. He believed that in the future city, the center should be accessible only to silent and slow transportation used for business. New, broad roads should be planned at the periphery, and factories at the remote corners of the city. "Today," Hellpach proclaimed, "the best assistant of the nerve specialist is the *engineer*" (Hellpach 1902: 38, my emphasis). Berlin's town-planning inspector G. Pinkenburg agreed. Although he did not share Hellpach's enthusiasm for the separation of work and home because it had caused an increase in transportion and thus of noise, he suggested new ways to construct pavements, cars, wagons, trams, tires, rails, and railroad crossings, in short, solutions that involved his own expertise, engineering (Pinkenburg 1903).

To these experts, technology was certainly not the crux of the problem. It was, as Pinkenburg added, the noise produced by man himself—shouting or the use of bicycle-bells and car horns—that was the most difficult to control. Therefore, man became the main focus of the first noise abatement campaigns. A few months after Lessing's *Der Lärm* appeared in print, the German Association for the Protection from Noise (*Deutscher Lärmschutzverband*) was founded. The Association became known generally as the Anti-Noise Society (*Antilärmverein*), and its journal was called Antirowdy (The Right to Silence). (Figure 4.1.) In New York, *The Society for the Suppression of Unnecessary Noise* was begun in 1906, and the London *Street-Noise-Abatement-Committee* was established two years later (Lentz 1994: 85).[3]

The rate of success of these early antinoise campaigns varied. Lessing's Anti-Noise Society was so successful it made the headlines of mainstream newspapers. It could boast of having over a thousand members, primarily scholars, physicians, lawyers, and artists. Among these were the writer Hugo von Hofmannsthal and the historian Karl Lamprecht (Vertrauensmänner 1910; Lentz 1998). The society presented lists of quiet hotels and attained some local success, such as the introduction of new pavements in specific streets, new ordinances for controlling ways of transporting goods, and limitating the use of train signals, steam hammers, and factory whistles (Birkefeld and Jung 1994). Yet, it did not

Der Antirüpel.

(Recht auf Stille.)

Monatsblätter zum Kampf gegen Lärm, Roheit und Unkultur im deutschen Wirtschafts-, Handels- und Verkehrsleben.

Organ des deutschen Antilärmvereins (Lärmschutzverband).

Herausgeber: Dr. Theodor Lessing, Privatdozent der Philosophie und Pädagogik an der techn. Hochschule Hannover. | Verlag der Aerztlichen Rundschau Otto Gmelin München ============ Liebherrstraße 8. ============

Figure 4.1 Front page of *Der Antirüpel (Recht auf Stille): Monatsblätter zum Kampf gegen Lärm, Roheit und Unkultur im deutschen Wirtschafts-, Handels- und Verkehrsleben* (Antirowdy [The Right to Silence]: Monthly on the Abatement of Noise, Rudeness and Uncivilized Behavior in German Economic, Business and Traffic Life), 2, no. 9 (September 1910). *Der Antirüpel* was the monthly of the *Deutscher Lärmschutzverband*, the German Anti-Noise Society.

achieve any changes in national law and was disbanded in 1914. Three years earlier, Lessing had left Germany for the United States. In the months before his departure, he expressed his disappointment about the failure of some members to pay their dues, and introduced his Berlin successors (Lessing 1911a; Lessing 1911b). His departure, however, seriously undermined the spirit and organizational strength of the society.

According to several historians, the lack of tangible success in the early German antinoise movement was due to the society's elitist point of view. "Tranquillity is distinguished" (*Ruhe ist Vornehm*) was its slogan. Indeed, articles in the society's journal persistently reproached the great mass of the people, working-class families and lower employees for making noise (Christ-Brenner 1910; Häsker 1910; Aus 1910; Perzynski 1911). Such an approach did little to foster alliances between the Anti-Noise Society and, for instance, labor unions. Moreover, Lessing's society had far less to say about industrial noise than about the noise of traffic, which again reflected its elitist approach. The press ridiculed Lessing and his supporters for their conviction that refined people were the most likely to suffer from loud sounds. Journalists even portrayed members of the

Anti-Noise Society as people fostering "a tyranny of the neurasthenics," practicing "a sport for intellectuals," and expressing a nonmasculine sensitivity (Auf 1910: 12–13, Ratschläge 1910: 33; Lentz 1998: 251). The outbreak of World War I only made matters worse; noise abatement campaigns now appeared completely insignificant by comparison (Birkefeld and Jung 1994; Lentz 1994; Saul 1996a; Saul 1996b).

The work of the environmental historian Raymond Wesley Smilor suggests that the early noise abatement campaigns in the United States were more successful. Just as in Europe, the predominately middle class American noise abaters associated silence with "civilization" (Morrow 1913: 121). Noise was not merely seen as primitive, however, but also, as we have seen, as inefficient in terms of the loss of productive power and deterioration of machines. What's more, noise in industry and business offices was widely believed to threaten employees' powers of concentration, and city noise in general was understood to undermine public health—a theme typical for the Progressive Era, in which urban reform was an important issue (Whiteclay Chambers II 1992; Melosi 1993). Urban reform indeed became the focus of the New York City Society for the Suppression of Unnecessary Noise, created by Julia Barnett Rice, a physician, mother of six children, and wife of the wealthy businessman, Isaac L. Rice. It recruited citizens with backgrounds similar to, but slightly more varied than the members of Lessing's society. By the end of 1907, membership in Rice's society had grown to about two hundred and included clergymen and businessmen (Smilor 1978: 66–67).

Instead of lightening the burden of noise on behalf of intellectuals, the New York Society aimed at reducing noise for the sake of children and the sick in order to promote processes of learning and recovery. It campaigned successfully for the creation of quiet zones around hospitals and schools, for enforcing laws against unnecessary horn signals in shipping, and for reducing the use of fireworks on the Fourth of July, which was both noisy and unsafe.[4] Their focus on the protection of children and the sick facilitated obtaining support from the press and industry, since—except in the case of the fireworks industry—no vested interests were at stake. The same was true for the focus on unnecessary noise, that is to say, noise that was not associated with technological progress. It

103

is also worth mentioning that the notion of quiet zones around hospitals and schools had its roots in the European islands-of-silence tradition, and was not, as Smilor claims, "an entirely new concept in urban planning" (Smilor 1978: 105). On the contrary, the fact that the idea of zoning *had* this history has probably facilitated its appropriation in the early-twentieth-century New York context.

Yet, Rice's antinoise society hoped for more than the regulation it had established just prior to World War I. The "unrestricted use" of automobile signals and muffler cut-outs remained to be tackled (Report [1914]: 20; Smilor 1971, 1978, 1980; Schwartz 1998).[5] The postwar heirs of the first-generation noise abatement societies would indeed pick up this challenge, as well as the New York society's extension of the city noise problem definition from a battle between barbaric and civilized people to a more general public health problem. Public education, however, was still seen as the crown of noise abatement, even after "objective" judgment had entered the scene.

"IN FIGURES PLAIN": GRASPING NOISE IN TERMS OF DECIBELS

Tackling traffic noise became *the* focus of noise abatement campaigns during the 1920s and 1930s. Street noise had acquired a very bad name because of its chaotic and irregular character—its lack of rhythm, at least to the ear of London's Noise Commission. Traffic noise was thus an obvious topic for the first noise measurements. These measurements re-enacted noise in terms of its loudness and highlighted traffic noise as the most predominant of all urban sounds. This only strengthened the noise abaters' obsession with the sounds of the streets.

Measuring the loudness of noise was not as self-evident as it may sound today. According to Birkefeld and Jung, the definition of noise as "unwanted sound" goes back as far as to the early Middle Ages, at least in general dictionaries (1994: 40). Yet within acoustics, "noise" was initially understood as nonperiodic, irregular vibrations in contrast to the periodic sound waves of musical tones. It was only in the 1920s and 1930s that the definition of noise as "unwanted sound" became common among acoustical engineers. Ronald Beyer claims that this meaning of noise arose when engineers working on telephone reception "found that the presence of other sounds interfered with such reception and began calling these extraneous sound noise," as well (Beyer 1999: 206).[6] I would add that the increasing dominance of English as the lingua franca in the academic world may have helped to establish this change. In German and Dutch, the terms *Gerausch* or *ruis,* meaning noise as opposed to "signal" are different from the terms *Lärm* or *lawaai,* meaning noise as in "clamor" (Wagner 1936b). The English language, however, uses only one word for both meanings.

At about the same time that the term noise became used by acoustical engineers, the measurement of sound drastically changed. The intensity of sound had been very difficult to measure because of the extreme low energy levels (expressed in watts) or levels of sound pressure (expressed in dynes per square centimeter) emitted. Only after the development of telephone technology, which enabled the separation of different frequencies (in Hertz or cycles per second), and of the radio valve, which made it possible to intensify small energy levels, sound intensity became easier to measure (Dubois 1937).

In noises, however, the range of intensities was so great that the use of the traditional physical units involved "almost astronomical magnitudes" that could "obscure the vital point of the effect of the noises on the ear." As became clear in experimental psychology during the late 1920s and the early 1930s, this effect was that the sensation level of sound, or loudness, depended approximately on the *logarithm* of the sound's intensity. This meant that the world of acoustics needed to measure noise "in some variety of sensation unit related logarithmically to the scale of physical intensities" (Free 1930: 19).

Such a unit for measuring sensation was borrowed from, again, telephone technology. In telephone technology, engineers had been searching for a standardized unit for measuring the transmission efficiency of telephone circuits, which resulted in the birth of the "bell," abbreviated as "bel," and of the "decibel" (dB) in 1925 (Beyer 1999: 222, 230; Martin 1929). This unit expressed a ratio of amplification: two sounds differed by 10 decibels, "when the louder is ten times the physical intensity of the fainter, by 20 decibels when the physical ratio between them is 100, by 30 decibels when the ratio is 1000, and so on" (Free 1930: 19; Davis 1931). Once they had a zero level of noise as the beginning of the scale, they could start counting. This zero became the threshold limit of a tone of 1000 Hertz (Hz).

———

Initially, American engineers employed the decibel only as the unit of noise. As early research comparing the loudness of pure tones indicated, however, the relation between sound intensity and loudness was logarithmical only at frequencies above seven hundred cycles per second. Tones of lower frequencies were perceived as relatively louder, which became visualized through B.A. Kingsbury's "equal loudness contours" (Kingsbury 1927). (Figure 4.2.) A few years later, Harvey Fletcher and W.A. Munson published additional data for frequencies above roughly five thousand cycles. According to them, these were perceived to be louder as well.[7] This discovery led the first International Committee on Acoustics (ICA), which met in Paris in 1937, to make the decision to work with two different noise units, the decibel and phon. It defined the decibel as the unit expressing the intensity of sound, and the phon, a unit of German origin, as the unit expressing the level of equivalent loudness. The noise to be measured had to be compared to a standard sound of one thousand cycles per second (or Hz), expressed in decibels above the threshold limit. This meant that if a noise was "judged by normal observers to be equal in loudness to the standard tone operating at a level of n decibels above [the threshold limit], the noise is said to have an 'equivalent loudness' or 'loudness level' of n phons" (Dadson 1949: 120).

At first the popular press foresaw a great future for the phon. A small poem in *The Observer*, signed by W.R., praised its appearance:

Hail! Newest unit, welcome to the host
　　Of ergs and amperes, kilowatts and therms,
Best of the lot, you shall be valued most
　　Among these unintelligible terms.

For you alone can make men realise,
　　In figures plain, the awful din they make,
So that at last some genius may devise
　　A means of curbing it, for Reason's sake.
　　(Quoted by Kaye 1937: 35)

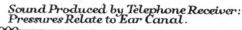

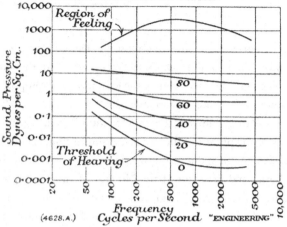

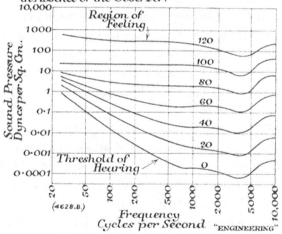

Figure 4.2 Equal loudness contours by B.A. Kingsbury (top),
and Harvey Fletcher and W.A. Munson (bottom), 1920s and
1930s. From Archives National Physical Laboratory, A.H. Davis,
The Measurement of Noise, paper read before Section G
of the British Association of Aberdeen, September 1934,
advance proof, [p. 1]. Courtesy National Physical Laboratory,
London. Courtesy of © Crown Copyright 1934.

Other units originally competing with the decibel were the "sone" and the "wien" (Kaye 1937: 35; Kerse 1975: 8; Wagner 1938). In the loudness scale based on the sone, the "number defining the loudness" was "proportional to the sensation of loudness" (Noise Measurement 1958/1955: 2–3). The idea behind the sone scale was to circumvent the counter-intuitive character of the logarithmical scales of the phon and the decibel. A similar unit, the wien, was proposed by Karl Willy Wagner, director of the *Heinrich-Hertz* Institute for Vibration Research at Berlin and president of the Noise Abatement Committee of the German Society of Mechanical Engineers. In a plea for the wien, he stressed that lay people and industrial managers had great difficulty understanding that a reduction from 100 phon to 80 phon involved a similar diminishment of loudness as a reduction from 40 phon to 20 phon. The public often assumed incorrectly that the first case implied a reduction of 20 percent and the second a reduction of 50 percent, while in both cases the newly attained levels were actually about one-tenth of the previous ones. Consequently, managers often felt unimpressed when engineers came up with reductions expressed in phons, no matter how unjustified. They believed that a proportional unit would be much more helpful to noise abatement than the phon and the decibel (Wagner 1938).

Notwithstanding such practical advantages of the sone and the wien over the decibel and the phon, the latter two units became the most commonly used in academic circles. Research institutions had taken the ICA's standardization of the units measuring noise seriously, and fostered the embedding of the phon and decibel in material practices, such as measuring instruments and graphs. Within the world of policy, however, talking about noise in terms of decibels eventually won out from expressing noise in phones. Most likely, a widely quoted review of noise surveys presented in decibels, published by Rogers H. Galt, a Bell Telephone Laboratories employee, may have influenced this outcome.

The first surveys of city noise, which had been carried out in New York, Chicago, Washington, D.C., and London, were published between 1926 and 1930. Initially, the newly available portable audiometers (or subjective noise meters) were used, but a short time later acousti-meters (also called objective noise meters) began to be employed.[8] The audiometer was picked up from medicine where it had been used to test people's hearing. It enabled one to measure the loudness of a tone by changing the intensity of a reference tone until it was felt to be masked by the tone being measured.

One version of the audiometer, the 3-A audiometer developed by Bell Telephone Laboratories, featured a buzzer that produced a standard noise covering "all frequencies over the usual noise range," while an attenuator was used to alter the amplitude of the noise it produced, and a telephone receiver to emit it through a hole in a face plate. This enabled the operator to listen to the standard noise and the noise entering from the surroundings simultaneously (Free 1930: 22; Fletcher 1929). In Europe, Heinrich Barkhausen developed a similar audiometer, while an audiometer made by the Western Electric Company used a microphone that picked up surrounding noise that was then compared to a noise standard. All audiometers, however, required operators who possessed extensive skills (Wagner 1931).

Still, the first New York City surveys, published in *Forum* magazine by Edward Elway Free in 1926 and 1928, were presented as "reasonably accurate and reproducible" (Free 1930: 23). The results showed, more than anything else, the loudness of traffic: "Trucks accounted for forty percent of the din, the elevated railway twenty-five percent, and surface streetcars twenty percent. The other fifteen percent arose from automobiles and horns, horse-drawn vehicles, fire apparatus, building operations, ambulances and police whistles" (Smilor 1978: 143).

The objective noise meters, also called "acoustimeter," "phonometer," or just "noise meter," were basically made up of a microphone, an amplifier, and an indicating meter, which made the measurement of loudness a purely physical issue. Furthermore, it was possible to "attach to the simple instrument a suitable electric network of condensers and inductances[,] the transmission characteristics of which for different frequencies approximate the sensitivity of the human ear at the same frequencies" (Free 1930: 27). In second half of the 1930s, this led to the use of noise meters—increasingly called as "sound level meters"—which could express levels of loudness for the higher (A) and the lower (B) frequencies. From the 1940s to the 1960s, national and international standardization of differentiations between dB(A), dB(B) and dB(C) followed suit (Martin 1991: 153).

In 1930, Rogers Galt published a review of all noise surveys in the *Journal of the Acoustical Society of America*. He demonstrated that the use of noise meters for measuring street noise was in principle not without problems. City noise consisted of an array of complex tones, whereas noise meters did not "sum up the components of a complex wave in the same manner as the ear in producing loudness. Hence, if two successive complex waves differ greatly in composition, the corresponding meter readings will not compare the waves as they are compared by the ear," since the meter results depended on the single frequency characteristics of the integrating device. Yet, because the New York City surveys showed that "the complex waves encountered in out-of-door noise survey are usually of the same general composition," different noise meter readings were thought to be comparable after all (Galt 1930: 33). Another problem he identified was that the noise meters could not trace exactly noises that fluctuated rapidly. Again, however, Galt considered this a minor problem with respect to traffic noise, for in the case of a passing elevated train, "the maximum or minimum [sound] is sustained over a period of two to five seconds" (Galt 1930: 35).

Galt's figures, graphs, and tables became widely known among both academics and noise abaters, and showed up in publications on noise in the United States as well as in Western Europe. (Figure 4.3.) Galt stressed that noise was not the same as annoyance, since it was also significant how often a particular noise occurred, which frequencies its components had, whether a noise was steady or intermittent, and whether the noise was considered to be necessary or not. Other early publications conveyed the same message. Despite such qualifications, however, noise levels increasingly became *the* sign of how bad the situation was, thereby objectifying the experience of sound in an unprecedented way.

The Need for a Noise Etiquette

By the late 1920s and early 1930s, big campaigns against noise succeeded the early noise abatement activities. In France, a Society for the Suppression of

NOISE LEVELS OUT OF DOORS DUE TO VARIOUS NOISE SOURCES					
SURVEY OF NEW YORK CITY NOISE ABATEMENT COMMISSION		NOISE LEVEL	OTHER SURVEYS		
DISTANCE FROM SOURCE	SOURCE OR DESCRIPTION OF NOISE		SOURCE OR DESCRIPTION OF NOISE		SURVEY NO.
FEET		DB			
		-130-	THRESHOLD OF PAINFUL SOUND		4
		-120-			
2	HAMMER BLOWS ON STEEL PLATE-SOUND ALMOST PAINFUL (INDOOR TEST)		AIRPLANE; MOTOR 1800 R.P.M.; 18 FT FROM PROPELLER		5
		110-	AERO ENGINE UNSILENCED — 10 FT		4
		-100-			
35	RIVETER				
15-20	ELEVATED ELECTRIC TRAIN ON OPEN STRUCTURE	90-	PNEUMATIC DRILL -10 FT.		4
			NOISIEST SPOT AT NIAGARA FALLS		2
15-75	VERY HEAVY STREET TRAFFIC WITH ELEVATED LINE	80-	HEAVY TRAFFIC WITH ELEVATED LINE, CHICAGO		7
15-50	AVERAGE MOTOR TRUCK		VERY NOISY STREET N.Y. OR CHICAGO		1
15-75	BUSY STREET TRAFFIC	70-	VERY BUSY TRAFFIC, LONDON		4
15-50	AVERAGE AUTOMOBILE				
3	ORDINARY CONVERSATION				
15-300	RATHER QUIET RESIDENTIAL STREET, AFTERNOON	60	AVERAGE SHOPPING ST, CHICAGO		8
			BUSY TRAFFIC, LONDON		4
15-50	QUIET AUTOMOBILE	50-	QUIET AUTOMOBILE, LONDON		4
	MINIMUM NOISE LEVELS ON STREET		QUIET ST BEHIND REGENT ST, LONDON		4
15-500	IN ENTIRE CITY DAY TIME / MIN. AVERAGE				
50-500	{ MIN. INSTANTANEOUS	40-			
50-500	IN MID-CITY NIGHT } MIN. INSTANTANEOUS				
		30-	QUIET ST, EVENING, NO TRAFFIC SUBURBAN LONDON		4
		20-	QUIET GARDEN, LONDON		4
			AVERAGE WHISPER -4 FT		3
		10-	QUIET WHISPER - 5 FT.		4
			RUSTLE OF LEAVES IN GENTLE BREEZE		3
		0	THRESHOLD OF HEARING		

Figure 4.3 Noise levels out of doors due to various noise sources. This figure was based on noise surveys executed in New York, Chicago, and London in the 1920s and produced by Rogers H. Galt (Bell Telephone Laboratories). From Rogers H. Galt, Results of Noise Surveys, Part I: Noise Out of Doors, *Journal of the Acoustical Society of America* 2, no. 1 (1930): 30–58, p. 38. Courtesy Acoustical Society of America.

Noise was established as early as in 1928. The German Society of Mechanical Engineers (*Verein Deutscher Ingenieure*) created a Noise Abatement Committee (*Fachausschuss für Lärmminderung*) in 1930. In the United Kingdom, the Anti-Noise League was founded in 1934, as were the Austrian Anti-Noise League (*Antilärm-Liga Oesterreichs*) and the Dutch Sound Foundation (*Geluidstichting*). The Dutch Anti-Noise League (*Anti-Lawaaibond*) followed in 1937. The British Anti-Noise League named their journal *Quiet,* which the Dutch imitated with *Stilte,* a journal first published in 1938. Local committees were also established, such as in Oxford and Amsterdam (Davis 1937: 132; Meyer and Potman 1987: 10–12; Agar 2002: 199; Horder 1936a: 5; Dubois 1938: 47; Noise 1935: 39–43; Zeller 1936: 10; Lee 1936: 18).

Just as with the first generation of antinoise organizations, these new noise abatement organizations did not become mass movements. As the British Anti-Noise League admitted in 1936, "The interest taken by the public in the cause of the League has been disproportionately greater than the growth of its membership roll" (Horder 1936a: 5). Press coverage of their activities was, nevertheless, substantial, especially in the 1930s, when hundreds of newspaper and magazine articles accompanied the campaigns. Moreover, noise abatement societies now generally viewed noise as a costly threat to the health and efficiency of *all* citizens. Henry J. Spooner, of the London Polytechnic School of Engineering, said that future generations would look back on an "age of folly vulgarized by an absence of quietude and repose, and notorious for uncontrolled devastating din that tortured the thinkers, deprived countless invalids and workers of recuperative sleep, impoverished owners of traffic route properties, increased the overhead costs in modern business and shortened the lives of countless sufferers" (Spooner cited in The World's 1928: 18).

Unlike their predecessors, the new noise abatement organizations could point out to sound levels in terms of decibels to underscore their sorrow about urban noise. The availability of graphs and figures did not immediately change the organizations' approach, however. Their leaders still focused on raising public consciousness. They used the noise measurements merely to illustrate the contribution of traffic to the problem of urban noise. Yet, as the noise abaters would soon find out, such measurements were not sufficient to influence the authorities' interventions. Nor did science always help to distinguish the right from the wrong sounds. Noise abaters therefore kept relying on former dramatizations of noise. Things would work out fine, they expected, if only the millions would follow them in their evaluations.

The efforts of the New York Noise Abatement Commission in the early 1930s exemplify how noise abatement organizations continued to approach the problem through public education. In 1929, in response to citizens' complaints, the New York Commissioner of Health appointed a commission to study city noise and find means of abating "the diabolical symphony" of "our present mechanical age" (Wynne 1930: 13). The committee's members were chosen from engineering, medicine, acoustics, the police and city administration, the automobile and telephone industries. It produced two reports, began a large-scale antinoise campaign, and considered its first report and the press responses it evoked to be important steps in creating public consciousness of noise. According to the first report, the noise nuisance varied from the use of loudspeakers outside shops, the screeching of brakes, and the abuse of automobile horns, to the use of muffler cut-outs on motor boats and the noises made by milk and ash cans, pneumatic drills, subway turnstiles, and elevated trains. A chart on the inside cover of the report provided an overview. (Figure 4.4.) The commission made clear that, up to the recent past, the noise of the machine-using age had proudly been perceived as the sound of progress and prosperity. Now, however, noise had to be considered as a serious health hazard (Brown et al. 1930: v–vii, 3, 57, 212).

The clamor of the city, the report asserted, impaired the hearing of New Yorkers and induced a harmful strain on the nervous system that led to "neurasthenic and psychasthenic states," to loss of efficiency of workers and thinkers, and to disturbed sleep (Brown et al. 1930: 17). In the late 1920s and early 1930s, experts all over the Western world echoed Josephine Goldmarks' previous claims about noise and fatigue in the textile trade (McKenzie 1928: 692–693; Wagner 1931). These claims were fueled by research on noise undertaken by John J.B. Morgan in the 1910s and by the industrial psychologist Donald A. Laird in

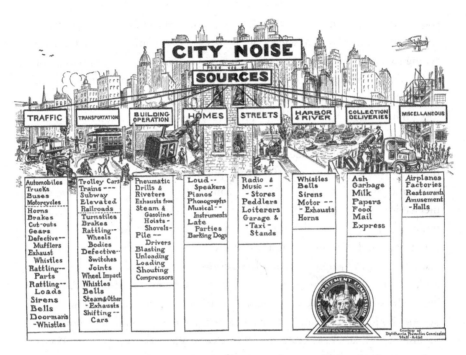

Figure 4.4 Chart on city noise published on the inside cover of Edward F. Brown et al., eds., *City Noise: The Report of the Commission Appointed by Dr. Shirley W. Wynne, Commissioner of Health, to Study Noise in New York City and to Develop Means of Abating It* (New York: Noise Abatement Commission, Department of Health, 1930).

the late 1920s. Morgan focused on energy expenditure, and studied the effect of noise on test subjects translating a code on a type writer. He concluded that noise raised the subjects' energy expenditure as they tried to overcome the auditory distractions (Smilor 1978: 145–146). Laird and his colleagues used a different approach. They confronted experimental subjects with unexpected noise, which produced more rapid breathing, raised systolic blood pressure, and reduced efficiency in response to the unexpected sounds (Sherman 1930; Laird 1930). They explained their results using a "fear reaction theory." Since noise had meant danger to "primitive man," it still caused a reaction of fear in his modern descendent (Smilor 1978: 17).

The New York Noise Abatement Commission contributed to the concerns about noise by referring to Morgan's work, reprinting one of Laird's scientific papers, and publishing a radio broadcast by the neurologist Foster Kennedy in their report (Brown et al. 1930). Kennedy had outlined the effects of sudden noises such as the burst of a "blown-up paper bag" on brain pressure (Brown et al. 1930: 109). "The noise raised the brain pressure to four times normal for seven seconds and kept it above normal for thirty seconds" (Smilor 1978: 231–232). Kennedy asked with obvious sense of drama, "If this is what an innocent paper bag explosion does to your brain, what does an unmuffled motor truck do to you?" Noise, thus, was "a definite menace to the public health" (Brown et al. 1930: 246–247).

The commission itself initiated a survey of noise based on 10,000 observations collected at ninety-seven outdoor locations in New York. A crew traveling on a truck specially equipped with an audiometer and a noise meter traveled over five hundred miles through the city collecting data. The survey's aim was not only to provide an overall impression of outdoor noise in numbers, but also to determine which sources contributed most to it (Brown et al. 1930: 111, 119, 32). Noise produced by traffic topped the list. It included the automobile horn (102 decibel), the motor truck (87 decibel), the streetcar (83 decibel), the noisy automobile (83 decibel), and the quiet one (65 decibel) (Brown et al. 1930: 112). At the same time, city newspapers published a questionnaire in order to map the roar of the city. Over eleven thousand people responded, and reported to be annoyed the most by the noise produced by traffic, transportation, and the radio (Brown et al. 1930: 26). Notwithstanding the public's irritation with the radio, commission members arranged radio broadcasts designed to "arouse public consciousness to the evils of noise and the advantages of a quieter city" (Brown et al. 1930: 74).

The commission reported some practical progress in noise abatement in New York, such as a reduction of unnecessary whistling in the harbor, of the blowing of car horns and of the use of open exhausts on mail delivery trucks, as well as the construction of more silent turnstiles. Moreover, it mentioned that the City Council had passed amendments to the Sanitary Code and the Code of Ordinances that controlled the use of loudspeakers. According to Dembe, the New York antinoise campaign also indirectly furthered the postwar financial recognition of industrial hearing loss by stimulating research and raising public consciousness on the issue (Dembe 1996). The second New York report, published in a limited edition in 1932, however, stressed the difficulty of solving the noise problem. Owing to the collapse of the U.S. economy, the commission was dissolved the same year (Smilor 1971: 35–36; Smilor 1980: 47–148).

The second New York report made clear that the commission could oversee research and propose methods of noise abatement, but had been unable to prompt city authorities to action beyond the few just mentioned. On almost every page, the report stressed, however, that the biggest issue was the public's consciousness of the issue. "The responsibility for city din rests less on the machine than on public apathy" (City 1932: 1). In a dramatic tone, the report told a story similar to the ones presented by Leo Marx's nineteenth-century literary men (discussed at the end of chapter 2), except for the last, crucial, turn:

One hundred and fifty years ago the world was like a quiet valley. . . . One day there was an ominous rumbling on the surrounding hills and a horde of barbarian machines poured down on the quiet valley. With steam whistle war whoops and the horrible clanking of iron jaws they came. . . . Resistance was vain, for how could hand power compete with steam? . . . But the swift of mind fled to the further hills where they tried to rebuild the lost world in their dreams. . . . Every aspect of the machine being considered marvellous, noise also became a minor god. . . . At last, however, the swift of mind returned from the ivory towers they had built in the hills and took over the task of recivilizing the new machine age. . . . But though many battles are won, the goal still lies far ahead. . . . We cannot expect quiet until the millions realise they have sold their birthright for a radio and an automobile. Therefore, the only permanent contribution the Noise Abatement Commission can make is to assist in educating the people. (City 1932: 2–4)

So, it was not a pastoral landscape that was sought for, but a quiet city (Thompson 1999). And it would be sought not by throwing out the radio and the automobile, but by changing public opinion. Even the law, in the form of standard noise ordinances, could be effective only if people understood that noise was unhealthy, inefficient, and often unsafe. Therefore, the police should act as an educational rather than a punitive body. Moreover, people should behave according to a Noise Etiquette for Automobile Owners. They should not blow their horns to summon people from inside their homes, but use them only when it was necessary to avoid accidents. Furthermore, they should buy a silent car, shift gears silently, have squeaking brakes realigned and the car greased regularly, and investigate unusual noises, making sure that the muffler was functioning properly and that all rattles were corrected.

Similarly, in Europe, noise campaigns focused on traffic and transportation, again with the automobile horn as the preferred object of criticism. Cities without a honking problem were considered rare, as a 1927 note by Walter Benjamin on Moscow reveals. "Of all the cities in the world, Moskow is the most silent, and when it is in snow, it is double silent. Here, the lead instrument in the orchestra of the street, the motor horn, has a weak position, there are few cars" (Benjamin 1980/1927: 97).

Research from across the Atlantic had indicated that, apart from loudness, horns emitting inharmonic overtones or sounds in either the lower or higher frequencies were the most objectionable (Brown et al. 1930: 161). In the United Kingdom, however, the issue was less straightforward. In 1929, the minister of transport, Wilfrid Ashley, convened three conferences on traffic noise for local authorities, police, and motoring organizations. Regulating the noise of clattering motor vehicles under the Motor Car Act appeared to be relatively easy: excessive noise included the sounds caused by "the faulty construction, condition, lack of repair," "adjustment," or "faulty packing" of the motor car (Report 1929: 2). Framing a definition that distinguished between horns with a "pleasing note" from those with "a strident and irritating one" proved far more complicated, however (Report 1929: 3).

At first, the conference participants tried to work out a regulation based on a distinction between the kind of horn used, which included a wind blown horn, an electric motor driven horn (a klaxon), and an electric buzzer horn. The National Physical Laboratory told them, however, "that it was possible to make horns of any of these groups with either objectionable notes or melodious notes." Other options were the use of headlights instead of horns in rural areas and at night, and the institution of zones of silence where motor horns "should not be permitted to be sounded at all" except when involving one's safety. The last suggestion "met with the uncompromising opposition of the motoring organizations." They accepted no interference with the drivers' discretion on this issue. In addition, they stressed that when there were accidents, police and insurance companies still asked whether drivers had sounded their horns or not, since they still saw sounding the horn as a significant means of preventing accidents (Report 1929: 3). The conference therefore suggested a further investigation into a "practical standard" that could be used. In the meantime, the public could be educated to avoid using a horn with disturbing notes, and not to use it unnecessarily. And since "a great deal of unnecessary horn-blowing in towns is due to traffic congestion" the conference participants underlined the importance of "adequate name-plating of streets." Such changes would reduce congestion, especially in London, caused by drivers who slowed down or even stopped to find out where they were (Report 1929: 4). Clearly, visualization had to enhance the smooth flow of traffic and thus to reduce noise.

This was not the end of the story, however. Publications such as *England, Ugliness and Noise* underlined that the "Juggernauts of the Road" and the noise of their motorized coaches threatened the peace of a large number of citizens, and—as the front-page suggested—of England itself (Darby and Hamilton 1930: 18). (Figure 4.5.) Lord Thomas Jeeves Horder, chair of the British Anti-Noise League, explained that citizens could "stand the blended sounds of traffic when these make a general hum." But "if, on top of this, an unsilenced car or motor-cycle accelerates down the street, not only our sense of hearing, but our sense of justice, is outraged" (Horder 1937: 11). His Anti-Noise League aimed at marshalling legislation and public opinion "to stamp out needless noise" (Horder 1937: 10; Purves-Stewart 1937). (Figure 4.6.) By such needless noise, he meant the sounds of inadequately silenced sports cars—the personal bête noir of Lord Horder—and honking (Horder 1936b: 7; Price 1934: 8).

———

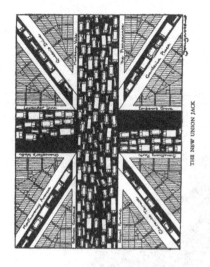

ENGLAND,
UGLINESS AND NOISE

By
AINSLIE DARBY and C. C. HAMILTON, B.A.

LONDON
P. S. KING & SON LTD.
ORCHARD HOUSE, WESTMINSTER
1930

Figure 4.5 Front page of Ainsly Darby and C.C. Hamilton, *England, Ugliness and Noise* (London: P.S. King & Son, 1930).

The league's efforts, as it and members of parliament proclaimed, helped to establish the 1934 amendment of the Road Traffic Act (Editorial Notes 1936: 3–4; Horder 1936b: 7; Strauss 1937: 19; Noise 1935: 7–8). This amendment prescribed a muffler that reduced the noise of the exhaust, and prohibited the sale and use of motor vehicles and trailers that produced excessive noise as a result of defects, lack of repair, faulty adjustment, or the faulty packing of a load—presumably the legacy of the transport conferences. Furthermore, the revised legislation banned the sounding of motor horns between 11:30 p.m. and 7 a.m. in built up areas. A few years later the Anti-Noise League also ensured that the Highway Code no longer contained "any positive recommendation to sound the horn" (Strauss 1937: 19; Strauss 1938: 13–14). It encouraged "those concerned in the design and the manufacture of noisy machines toward better manners as well as more efficient science," and organized a Noise Abatement

Figure 4.6 Advertisement of the British Anti-Noise League, 1935. From *Noise Abatement Exhibition* (London: The Anti-Noise League, 1935), p. 91.

Exhibition aimed at familiarizing the public with artifacts varying from noiseless typewriters and ear plugs to quietly running electric motors, silenced breakers, exhaust silencers, and pneumatic railcars (Noise 1935: Foreword). Revealingly, one of the advertisements in the exhibition catalogue stated that road drills could "be effectively silenced without loss of power" (Noise 1935: 79). (Figure 4.7.)

In the late 1930s, the league often stressed that road drills, trains, and cars—"when properly driven"—had indeed become quieter (Longer 1937; Prioleau 1938; A Nearly 1938; Quiet 1938). At the same time, maximum loudness levels reached the public agenda. In 1937, a Departmental Committee chaired by the physicist George William Clarkson Kaye[9] recommended that the noise of a new vehicle should not exceed 90 phon. For used vehicles the maximum should be 95 phon, and for warning devices 100 phon. Research into the transition point between tolerable and excessive vehicle noise had inspired these limits (Kaye and Dadson 1939: 731, 756). Physicists were still unable to sort out the difference between agreeable and disagreeable sounding horns, however. A jury test involving 200 listeners revealed "wide divergences of opinion on the question of annoyance." Yet loudness was "a factor of first and practical importance" (Kaye and Dadson 1939: 750, 751). Such findings increased the options for intervention. Nevertheless, what must still be done, the league pointed out, was to spread "a feeling that to make a needless noise is bad manners. We are educational rather than condemnatory" (The Annual 1938a: 30). Or, in terms of the New York campaign, there was still a need for a noise etiquette.

Paris authorities prohibited the blowing of horns after the French Society for the Suppression of Noise addressed the Prefect of Police on the issue. Also, an educational campaign by the society had to improve "the jay-walking habits of pedestrians' (Brown 1930 et al.: 10; Contre 1936: 27). (Figure 4.8.) Italy banned the use of the horn at night in 1934 and, from 1935 onward, extended their prohibition during the day in Rome. Budapest prohibited honking in one of its main streets and in 1937, Bukarest restricted horn blowing at night. By the same year, Germany had made mufflers obligatory, confining the sound of engines to a maximum loudness of 85 phon and that of horns to 100 phon. Moreover, horns had to have an "approved pattern" and "sound of unvarying pitch," and

That
ROAD DRILLS
can be
effectively silenced
without loss of power

is proved by the decision of
The Westminster City Council

immediately put into effect, as shown by
the illustration above (reproduced from a
photograph taken on 9th April, 1935, at
Grosvenor Road, Westminster).

Contractors: The Improved Wood
Pavement Company, Ltd.

BROS. LTD.

CAMBORNE - ENGLAND

LONDON OFFICE · · · BROAD STREET HOUSE, E.C.2

Figure 4.7 Advertisement by Holman Bros. Ltd.,
Camborne, England, 1935. From *Noise Abatement
Exhibition* (London: The Anti-Noise League, 1935),
p. 79.

CONTRE LE BRUIT

Peuple de Paris, Reclamez le silence !

Si un pieton courait sur le trottoir en poussant d'effroyables hurlements, qu'arriverait-il ? Immediatement empoigne par un agent, il serait traine au poste pour infraction aux bonnes mœurs. Cependant sa conduite ne serait pas aussi reprehensible que celle d'un conducteur d'automobiles, qui fait un raffut du diable avec son klaxon ou sa trompette.

Si vous tenez a avertir les pietons, Messieurs les automobilistes, ayez la bonte de corner avec plus de douceur. Pourquoi ameuter tout un quartier ? Il n'est pas necessaire d'ecorcher les oreilles, de lacerer les nerfs.

Nous demandons pour Paris plus de silence. Le tintamarre epouvantable dont nous assourdissent les chauffeurs, les mecaniciens de chemins de fer et de bateaux, ne donne pas une haute idee de notre civilisation. Le bruit detruit tout repos; il empoisonne la vie, nuit a la sante. A quoi bon nos beaux monuments, nos sites merveilleux. nos jardins delicieux, si le visiteur affole, abruti par ce chaos de rumeurs, ne peut pas en jouir ?

Les autorites municipales auxquelles se joignent les bons citoyens, veulent le calme. Aidons-les de notre mieux ! Reclamons tous plus de silence !

La Societe pour la suppression du bruit
10, Rue de l'Elysee, Paris.

Imp. SCHNEIDER, 113, Av. de Roche, NEUILLY

Figure 4.8 Postcard issued by the French Society for the Suppression of Noise, saying: "Against Noise: People of Paris, Reclaim Silence!" 1936. From Contre le bruit, Quiet 1, no. 3 (Autumn 1936): 27.

the use of horns for other purposes than warning was punishable (Promoting 1937: 31; Promoting 1938a: 35–36; Spitta 1941: 25; Trendelenburg 1935: 142). (Figure 4.9.) And cities as diverse as Prague, Wiesbaden, Stuttgart, Zurich, Stockholm, Milan, Rome, Antwerp, Brussels, and The Hague organized "days of quiet," "silence weeks," and "traffic weeks" to educate the public in reducing the deafening use of car horns (Breda bindt 1935; Meer 1935; Algemeene 1936: 5; Verslag 1934: 33; Verslag 1936: 59–78; Promoting 1938b: 33).

These campaigns did indeed create a different city sound. By reducing the use of car horns, creating traffic control, and stimulating city planning, the silence campaigns contributed to a new rhythm of city life. As historian Anthony McElligott stated in a recent, highly original interpretation of Walter Ruttmann's film *Berlin: Symphony of a City* (1927), the spectacular image of the fast flowing traffic "might appear chaotic, but it never is. Instead its constituent and apparently anarchic parts are constantly configured into a single pulsating flow of order as a result of an imperceptible traffic plan, and the regulating presence of the traffic police, who together ensure that the underlying structure of urban-based capitalism remains intact" (McElligott 1999: 210). *How* such a flow of order was reached in the Netherlands, is the topic of the next section. As the British example makes clear, the regulation of honking was not immediately welcomed by motoring organizations. What, then, contributed to its acceptance? And what did this mean for the reconceptualization of the traffic noise problem generally? The latter question will be the subject of the conclusion.

SILENCE CAMPAIGNS: CREATING AN URBAN RHYTHM

In Holland, it was scientists and engineers who took the official first steps toward studying and fighting the noise problem.[10] The first president of the Dutch Sound Foundation, established in 1934, was Adriaan Daniël Fokker, a professor in technical physics who had ventured into music in the second half of his career.[11] The aim of the Sound Foundation was to spread knowledge about and intervene in all kinds of sound issues. Since noise had a bad influence on man's ability to work, society was in need of scientific research that could lead to suggestions for maximums of allowed sound levels, and for noise reduction of artifacts such as vacuum cleaners. Designing prescriptions for how to build housing, inspecting and measuring construction materials, providing noncommercial advice, and public education were the Sound Foundation's other tasks.[12] (Figure 4.9.) Engineers' interest in such activities may partially have stemmed from concern about the insecure job market at the time.[13] Only several years after its formation, the Foundation raised a separate Anti-Noise League. Its goal was to support local noise abatement committees and to help the Foundation financially, thus enabling it to focus on research.[14]

In both 1934 and 1936, the Sound Foundation organized Anti-Noise Conferences in cooperation with the Royal Dutch Automobile Club (*Koninklijke Nederlandsche Automobiel Club*) (Verslag 1934; Verslag 1936). According to Fokker, conferences such as these were a complaint against modern culture. To fight the "demon" of noise, one had to set out "the ideal of the expert professional who silently knew to control his noiseless machine" against the "noise vulgarian" and "motor yokel" who tried to impress others by making noise (Verslag 1934: 16). The despised blowing of car horns, most contributors made clear, had to be eliminated first. Car owners were used to blowing their horns at every side street, and expected the question "Did you blow?" whenever they were involved in an accident. A ban on acoustical signals at night could encourage car drivers to slow down and signal optically by using their head lights. Similarly, pedestrians and cyclists should learn to behave correctly. Pedestrians must cross intersections in a straight path, and cyclists should stay to the right side of the road. If such measures were adopted nation-wide, they would reduce the number of accidents, as had been proven abroad. Thus, if society replaced aural forms of traffic control by visual and spatial ones, silence, safety, and order, it was believed, would follow.

Quiet streets could also be realized by new spatial designs. City and traffic noise, one conference participant claimed, were almost "directly proportional to the age of the city. . . . The narrower the streets, the more winding their plans and the more unsurveyable their angles, the bigger . . . the disturbance of traffic noise" (Verslag 1936: 17). In the old parts of cities, noise-producing industries

PREFECTURA POLIȚIEI MUNICIPIULUI BUCUREȘTI
SERVICIUL CIRCULAȚIEI

Conducători de automobile, amatori, șoferi și watmani de la tramvaie!

Liniște în timpul nopței Circulați fără a întrebuința clacson sau clopot

Semnalizați la nevoie numai prin lumină.

Cetățenii Capitalei au nevoie de liniște în timpul nopței, și de aceia vă cerem să ascultați de dispozițiile ordonanței No. 318 din 22 Septembrie 1931 a Prefecturei de Poliție.

Conducătorii și vatmanii cari vor contraveni acestor dispozițiuni vor fi amendați pe loc cu amendă până la 1000 lei și ridicarea permisului de conducere, în afară de aplicarea dispozițiunilor art. 587 al. 5 din noul cod penal, care prevede închisoare polițienească de la una la 10 zile și amendă de la 100 la 1000 lei, pentru conducătorul de automobil care, în contra regulamentului de circulație sau fără o justă necesitate, prin instrumente sonore sau prin manevrarea motorului, face astfel de sgomot încât turbură liniștea locuitorilor.

Șeful Serviciului Circulației

Chestor, GEORGE BOTEZ

1 Iulie 1937.

Tipărit și distribuit gratuit prin Prefectura Poliției București.

C. 33'5. Tip. Pref. Poliției.

Figure 4.9 Notice issued by the Bucharest Police, 1938. It addressed the drivers of motor cars and tramways, and required them to drive without honking or the ringing of bells during the night. From Promoting Quiet Abroad, *Quiet* 2, no. 7 (January 1938): 34–36, p. 36.

were loosely scattered over the city area. Obsolete means of transportation, such as the electric tram, blocked normal traffic. Therefore, town plans needed modernization. The separation of industrial and residential areas, a restraint on "ribbon" development (narrow built-up areas along roads), and the reduction of angles and bends in roads, were of the utmost importance. The conference participants also discussed improvements in the sound levels of technological artifacts themselves, for instance by creating mufflers, new horns, and alternative ways of loading. were also discussed. Yet, it was not these issues that took center stage at the antinoise conferences. Instead, the production of orderly town plans and, above all, orderly behavior took priority. As to perfect conduct, this should be achieved by orchestrating silence campaigns.

In 1935 and 1936, noise abatement committees organized silence campaigns in Breda, The Hague, Rotterdam, Groningen, and the province of Limburg, often in cooperation with the police and traffic organizations. (Figure 4.10.) During "silence weeks," "silence months," and "silence exhibitions," the campaigners distributed thousands of pamphlets, placards, and flags. Dozens of newspaper articles, radio broadcasts, and even newsreels shown in cinemas covered the campaigns. The basic consideration was to familiarize civilians with the idea that they had to *look out* before they blew their horns or before they forced others to blow their horns. "Use your eyes in stead of your horn," one pamphlet advised. Just as railway stations had replaced the infernal noise of bells and whistles by optical signals, streets should likewise become quiet. People should watch out, stay right, slow down, use mufflers, and they were summoned to consider their motor as a means of transportation, instead of a machine for testing other people's eardrums. (Figure 4.11.) The back page of one pamphlet even advertised a prize contest with questions on the proper traffic behavior, which was later won by a musician who stressed that all traffic participators could contribute to decreasing the din of horn blowing.[15] As one of the campaign's slogans suggested, "Orderly Traffic promotes Silence" (Promoting 1937: 32).

So, chaos meant noise and order meant silence—conceptual combinations with a long history, as we have seen. Yet the focus on traffic behavior and control did not only stem from the legacy of earlier definitions and dramatizations of

––––

Figure 4.10 Leaflet "Anti-Noise Week," September 23–28, 1935, The Hague (The Netherlands). From Archives Sound Foundation (*Geluidstichting*), stored at the archives of the Dutch Acoustical Society (*Nederlands Akoestisch Genootschap*), Nieuwegein, The Netherlands. Courtesy Nederlands Akoestisch Genootschap.

Voorkomt ONNOODIG LAWAAI

Wielrijders,

voorkomt door U te houden aan de verkeersregels, dat anderen U met lawaai aan Uw plichten moeten herinneren.

Motorrijders,

bedenkt, dat Uw machine een vervoermiddel is en geen apparaat om de trommelvliezen van anderen op sterkte te keuren. Zorgt derhalve voor een doelmatigen knaldemper.

Automobilisten,

vermijdt zooveel mogelijk het hinderlijk of onnoodig signaal geven; mindert liever Uw vaart, dit is veilig en niet hinderlijk.

Eigenaars en chauffeurs van vrachtauto's,

stuwt Uw vrachten goed; dit voorkomt lawaai en spaart de lading.

Figure 4.11 "Prevent Unnecessary Noise," second page from the Leaflet "Anti-Noise Week," September 23–28, 1935, The Hague, The Netherlands. From Archives Sound Foundation (*Geluidstichting*) (1933–1942), stored at the archives of the Dutch Acoustical Society (*Nederlands Akoestisch Genootschap*), Nieuwegein, The Netherlands. Courtesy Nederlands Akoestisch Genootschap.

noise. Behind the emphasis on traffic education stood a fascinating *discourse coalition:* a coalition enabled by the "discursive glue" of a particular phrasing of the problem rather than by "shared material interests" (Lees 2004: 102). Initially, the Sound Foundation had hesitated in deciding which topic was most appropriate for a public campaign. In any case, it had to be tangible in order to raise public attention. By chance, the Sound Foundation found out that the Royal Automobile Club had plans for orchestrating silence weeks. To the Automobile Club, such events represented a chance to strengthen their competitive position with respect to the General Dutch League of Cyclists (*Algemene Nederlandse Wielrijders Bond*). At the same time, silence weeks offered an opportunity to promote the automobile in general, since orderly behavior of all traffic participants literally cleared the way for the car. What's more, cooperation between the Royal Automobile Club and the Sound Foundation meant the financial backing of the latter by the former. Such factors facilitated the choice to campaign against traffic noise.[16] The dual meaning of the slogan "Orderly traffic promotes silence" helped to sustain this coalition: it referred to cleared and, therefore, honk-free streets (the interest of the automobile owners) as well as to a reduction of the fatiguing chaos of sounds (the aim of the noise abaters). (Figure 4.12.) This dramatization of noise linked two different, if not opposing, definitions of the street noise problem into one approach to noise, and enabled the coalition at hand. Such a multi-interpretability of key notions, other studies have shown, is often typical of processes in which language alters the perception of issues and leads to successful discourse coalitions (Lees 2004, Hajer 1995).

To the third party in this connection, the police, the educational approach may have been particularly attractive. As historian Uwe Fraunholz has shown for Germany, policemen and civil authorities dealt more gently with car drivers as soon as they became drivers themselves. Instead of punishing unwelcome behavior, they increasingly responded with "discrete admonitions," enabling an "unimpeded circulation of cars" (Fraunholz 2003: 117, 130). The discourse coalition for public education expanded even wider. In 1935, the national-socialist government became involved. It financed a federal campaign for a "noiseless" week in response to the Führer's instruction to secure "the strong nerves" of the

Figure 4.12   Some Noise Abatement Adherers behind Some Sources
of Noise, The Netherlands, mid-1930s. The second and third persons
from the right are C. Zwikker, and A.D. Fokker respectively. A. de Bruijn,
50 jaar akoestiek in Nederland (Delft: Nederlands Akoestisch Genootschap,
1984), p. 9. Courtesy Nederlands Akoestisch Genootschap.

German nation. The "victory in the struggle for existence," so the one campaign
pamphlet said, would "always be for the nation with the strongest nerves." Here,
the driving thought behind noise abatement clearly involved more than restoring
public health by reducing the fatiguing chaos of sounds. Noise abatement had to
make a nation win in the struggle with other nations (Wagner 1935: 531).

The Dutch silence campaigns indeed succeeded—at least temporarily
and locally—in diminishing the use of car horns. In The Hague, for instance,
the police counted the number of horn blows on several days at three locations at fixed hours of the day. Before the silence week, the 5,300 buses, cars,
trucks, and motors that passed produced over 5,400 horn signals. A month
after silence week, over 5,100 vehicles passed and only slightly more than 2,500
horns were sounded. No comparable figures are available for Rotterdam, but
observers reported "progress" and a notable reduction of noise at particular

places (Verslag 1936: 24, 30). Furthermore, the national government renewed the Motor and Bicycle Regulation in 1937, which included maximum sound levels for motors and horns, and a ban on using horns at night. In 1934, a British industrial psychologist claimed that a similar clause in his country's legislation had been "spontaneously followed" by a "striking reduction in the frequency of the hoot of the motor-car horn during the day." The Noise Abatement League confirmed this a few years later (Editorial Notes 1938: 6). And the Stuttgart police reported "a general calming of the traffic," as well as a relative "drop of the number of casualties" (Myers 1934: x; McElligott 1999: 222; Price 1934). In Breda, the campaign had less success. According to a member of the Sound Foundation, this was due partly to Breda's narrow streets. A newspaper, however, stressed that policemen just outside Breda kept instructing motorists to sound their horns at every crossing (Breda 1935; Verkeerszondaars 1935).

Notwithstanding such problems, the use of the motor car horn seems to have been reduced generally. Other changes are also worth mentioning. According to one of the Dutch Anti-Noise Conference reports, the behavior of pedestrians did not improve, but that of bicyclists, motorcyclists, and motorists did (Verslag 1936: 26). In Holland and elsewhere, traffic acquired a more visual character through the use of headlights and optical train signals (Dembe 1996: 172). And as McElligott has documented for Berlin, city traffic came to be "collectively directed by traffic signs, regulations, and police hand signals." Commuters circulated by rail, tram, and roads, "movements forming a grid that contained and controlled the energies of the metropolis" (McElligott 1999: 223). Such changes fostered economic well-being and safety, and at the same time created a rhythm out of chaos, thus reasserting the feel for human control over events, in a way that is comparable to the rituals described by anthropologists for more ancient or remote cultures. Transformations in pavements, from cobblestones to asphalt, for instance, added to the new smoothness (McShane 1979). Yet the *form* of control was new, and was part of what historian of technology Thomas P. Hughes has characterized as the "values of order, system, and control" that engineers embedded in machines, and that "have become the values of modern technological culture" (Hughes 1989: 4).

Dutch antinoise movement activities that began in the late 1930s, such as those concerning noisy dwellings and the noise of trams, aircraft, and "stamping machines in postoffices" were far less successful than the campaigns fighting traffic noise. Their lack of success was due to a lack of organizations with which associate their campaigns, the economic crisis, and, finally, to the war. In the United Kingdom, a newspaper ridiculed the Noise Abatement League by publishing a fictional plan to "deaden the sound of falling bombs in time of war." The League dropped its first campaign against aircraft because it was believed "unsuitable," given "the need for re-armament" (Mr Punch 1938: 14; The Annual 1938b: 24). In 1940, the Dutch Anti-Noise League stopped all its activities.[17] After the war, a new noise abatement organization was created in its stead, but slumbered until a revival of the noise abatement movement in the 1970s.[18]

CONCLUSIONS

The many manifestations of technology in the late nineteenth and early twentieth century drastically changed the sonic environment in Western cities. The sounds of factories, trains, trams, automobiles, buses, motorcycles, telephones, gramophones, radios, and steam hammers accompanied those of church bells, whips, street musicians, carpet beating, milk cans, and yelling people.

By articulating the "city noise problem" and the proposed solutions to it out of the technological sounds mentioned above, antinoise campaigners employed many of the older repertoires for dramatizing sound identified in chapter 2, as did the people responding to the campaigns. Against tranquility as the sign of intellectuality and prominence stood noise as a primitive form of self-diffusion and raucous-mindedness. This age-old association between silence and social distinction guided the early German campaigns, which focused on creating silence for the bourgeois elite living in densely populated areas. Despite the many practical proposals these campaigns advocated, promoting the respectability of silence was seen as crucial for noise abatement, whereas industrial noise was not given any priority. This hampered alliances with labor unions and inspired opponents to counter the antinoise campaigns with another association: that between noise sensitivity and hysterical femininity.

Whereas the city noise problem as a whole was phrased in terms of civilization versus barbarism, and thus as a threat to social order, street noise was seen as a particular form of disorder. Its irregular and chaotic sounds lacked rhythm and thus posed a danger to mankind—again a long-standing dramatization of noise. Indeed, turn-of-the-century city streets were filled with a large variety of sounds from pedestrians, bicycles, horses, and carts to streetcars, elevated trains, and automobiles. Auditory signaling, often triggered by past regulations, was highly common. That traffic noise and signaling became the focus of city noise abatement was not self-evident, however. As a *New York Times Magazine* article about the London campaigns subtly revealed, honking was only "a small part" of the chorus of urban sounds (Price 1934: 8). Yet the blowing of horns perfectly fit into the definition of the noise problem as a fatiguing chaos of sudden sounds resulting from bad manners. Honking, the screeching of breaks, and the sounds of the unmuffled sports car were the logical successors of vendors' cries, whip-cracks, news boys' shrieks, and ship signaling. The general discourse on the nervousness of city life and urban reform, and the individual ownership of automobiles and gramophones provide even more context for understanding the obsessions of the city noise abaters.

The result was a consistent focus on public education. The noise abating associations were antimodern, yet not antitechnology, and combined proposals for technical solutions with a deep faith in and hope for a general noise consciousness. Remarkably, neither the increasing concern with noise as a general health hazard that affected individuals' efficiency negatively, nor the introduction of measurement devices altered antinoise organizations' rhetorical focus. Indeed, the loudness of traffic noise came to the fore. Yet measuring noise in decibels did not immediately help the noise abatement associations in prompting authorities into action. Instead, teaching the public a noise etiquette remained, more than anything else, the focus of noise abatement. After all, noise abaters made clear, those who did not silently master their machines displayed vulgarity.

It was the symbolic link between silence and order that enabled the notable discourse coalition between motoring organizations, noise abating societies, and the police. The regulation of traffic, and the education of the public about these rules had to bring silence, safety, and orderly streets. Silence campaigns indeed successfully reduced honking, as other kinds of signaling had been successfully restricted. At the same time, the campaigns visualized city life: by promoting the use of headlights instead of horns and by fostering name-plating, traffic lights, pedestrian crossings, notice signs, and visual announcements. Visualization was thus the consequence of noise abatement rather than the cause of the neglect of noise problems. Together with spatial solutions, such as enlarging and straight-ening streets and the introduction of asphalt, rubber tires, and new rails, traffic control and education generated a new urban rhythm. This rhythm successfully tackled the noise problem in terms of its old definition: the chaos of sounds. City noise became the general hum of blended sounds that Lord Horder had always preferred.

It was only after the establishment of this new smoothness of traffic that the measurement of noise could really contribute to a new conception of the noise problem. After signaling had been subdued, street noise became more amenable to measuring. To understand this, we have to return to Galt's review of noise surveys once more. In his view, the noise meter could grasp street noise despite the difficulties in measuring complex and rapidly fluctuating noise. At the same time, however, he clarified that his claims about the relationships between sound intensity and the number of vehicles passing by only applied to homogeneous traffic, and not to locations where other loud sounds were present (Galt 1930: 50). The urban noise of the late 1930s was therefore more fit for the new measuring instruments than before. Now that the car had become more widely used, but horn signaling and sputtering exhausts less common, the defi-nition of the noise problem could change from a *chaos* of sounds coming from *individual* noise sources to *levels* of sound produced by a *collective* source.

Assessed in terms of the criteria suiting this *new* definition—reducing levels of noise—the strategies of the noise abating societies failed. Their leaders, as Smilor (1971) has stressed, tragically overlooked the quantitative increase in traffic. And as Emily Thompson has claimed for the United States, the success of the suppression of noise through "sound absorbing building materials" that had been provided for by acousticians and used to transform "homes, offices, hospitals, and hotels" into "shelters" from noise "may be partially responsible for the failure" of the noise abatement campaigns (Thompson 2002: 167–168). We should not overlook, however, that the silence campaigns in Europe seriously restricted the blowing of horns and other sudden sounds, and were thus a success in terms of reducing the chaos of sounds.

The social elite's pride in its refined sensibility—its subjective perception of sound—was helpful in this process, although it blocked coalitions with labor organizations. Yet the elite defined which sounds were most offensive. They succeeded in disciplining the masses to refrain from making at least the most striking noises, the most brutal to the elite ear, thus partially controlling which sounds were allowed and which were not. This would change dramatically, however, as chapter six on the noise of neighbors will elucidate. In postwar Europe, the cheap residential housing from the reconstruction era was less well sheltered from noise than American skyscrapers, and the noise of neighbors became a fiercely discussed topic. It was only in these debates that the subjectivity of sound perception acquired a meaning that fundamentally hampered noise abatement. And as new variations on old songs in praise of noise—variations that celebrated its positive connotations—entered the world, alternative noise abatement strategies were asked for.

One of the press responses to the New York campaign pointed to such a positive connotation of noise: "Isn't it precisely . . . the big noise, the detonation of our national dynamite, that attracts the big crowds which make New York?" (City 1932: chapter 8, 18). And in London, an anonymous commentator on the first annual report of the London Anti-Noise League claimed that the noise of machines would always "find out a way of returning," since the "joy of life expresses itself in a crowded chorus" (Y.Y. 1935: 274). It was as if these journalists had hired Italian futurists to write their articles.

5

THE ART OF NOISES:
THE CELEBRATION AND CONTROL OF MECHANICAL SOUND IN MUSIC

RIOTOUS CONCERTS AND COUNTERPOINT HISTORY

In 1927, George Antheil's musical composition *Ballet Mécanique* made its American premiere in Carnegie Hall, featuring ten pianos, a player piano, xylophones, electric bells, sirens, airplane-propellers and percussion. Staging an aircraft engine as a classical musical instrument was not exactly common practice. The New York premiere had been advertised by referring to the riotous reception of the first performances of *Ballet Mécanique* in Paris the year before, and Antheil himself had been described as a "sensational American modernist composer." However, the New York audience and critics expressed a deep disappointment. They neither appreciated the ballet's pompous publicity or its performance. It was a "Mountain of Noise out of an Antheil" went one headline. "Boos Greet Antheil Ballet of Machines" said another (Whitesitt 1983: 31, 37).

The concert as well as the rumors that surrounded it had a lot in common with the first public performance of Luigi Russolo's compositions *Awakening of a City* and *Meeting of Automobiles and Airplanes* in Milan in 1914. (Figure 5.1.) Russolo's concert had followed on the heels of the publication of his futurist manifesto, *The Art of Noises,* in which Russolo had proclaimed a renewal of musical compositions by enlarging and enriching the limited timbres of traditional

Figure 5.1 Sample from the score "Risveglio di una città"
(Awakening of a City), Italy, 1914, composed by Luigi Russolo.
From Luigi Russolo, *The Art of Noises* (New York: Pendragon Press,
1986/1916), p. 72.

orchestras with the infinite variety of timbres of noises. For his compositions, or "networks of noises," Russolo used a complete orchestra of newly invented noise instruments. In 1914, twenty-three such instruments were on stage. A huge crowd had gathered, whistling, howling and throwing things even before the concert had started, and the audience remained uproarious throughout the performance (Russolo 1986/1916: 34).

In the 1910s and 1920s, riotous concerts such as Antheil's and Russolo's were not unusual. The premiere of Igor Stravinsky's *Sacre du Printemps* in Paris in 1913 caused such noisy protests that the concert visitors' sounds became part of the performance. The same year, Arnold Schoenberg's *Kammersymphonie* met with shouting and the rattling of keys from the audience in Vienna. Nor was the representation of machines and machinery noise in music unusual. In 1846, Hector Berlioz composed *The Singing of the Train* for choir and orchestra, and Arthur Honneger presented his *Pacific 231,* an orchestral work about a steam locomotive, in 1923. Even the use of mechanical instruments, such as the player piano was far from new. Compositions had been written for automata as early as in the eighteenth century, by Mozart for instance (Kowar 1996). A real novelty, however, was the construction of music out of the everyday noise of machines by using machines or machine-like instruments instead of traditional orchestral instruments. Many of these instruments were invented in the first decades of the twentieth century. Some were automata; others used electricity and radio waves to create new timbres and tones. (Braun 1992; Kahn 1992; Braun 1994; Weidenaar 1995; Chadabe 1997; Théberge 1997; Kahn 1999; Glinsky 2000; Braun 2002; Pinch and Trocco 2002).

The introduction of machines and mechanical noise into musical compositions has often been discussed as the legacy of the futurist movement and its fascination with machine sounds in everyday life. Futurist artists indeed adored urban noise, which they viewed as the symbol of a new and thrilling mechanical age. To them, as to many urban dwellers, noise was *the* attraction of the metropole, a sign of the exciting speed of life. At first sight, the connection between music-made-out-of-mechanical-sound and the celebration of city noise seems to be the best argument for writing about music in a book about noise. In such an account, the champions of the art of noise would be cast as the most fierce opponents of the noise abatement organizations.

On closer analysis, however, the relation between the art of noise and the history of noise abatement is a more interesting one. The best reason for including art music in a book on noise is not situated in the differences between a movement embodying a turbulent love of noise versus a movement with a deep hatred of noise, but in their less obvious similarities as two forms of activism. To put it bluntly, in the first decades of the twentieth century, both noise abaters and noise artists strived to attain a genuinely elitist control of sound. In this sense, the art of noise was less "modern" and more "romantic" than the love of machines initially might suggest. The machine, as avant-garde composers and critics stressed, enabled the direct expression of a composer's creativity. By using a machine's mechanics, sound could be tamed, molded, and infinitely transformed as a unmediated representation of the artist's mind. Moreover, machine sounds helped to evoke the quintessence of life, another fundamentally romantic aspect of machine art.

Moreover, the noise abaters' and the noise artists' movements did not only resemble each other in their aim for an elitist control of sound, but also in their fate. Their leaders generated a lot of publicity and gained some striking, yet limited successes, at least—and this is a significant qualification for music—in the concert hall. For this reason, this chapter serves as a counterpoint to the rest of the book. Like counterpoint voices in music, the noise artists both opposed and echoed the obsessions of noise abatement organizations' leaders. The ultimate aim of counterpoint, both in music and in this book, is heightened movement. Some notes point backward, others forward, yet the combined result is motion. In the first decades of the twentieth century, both noise abaters and noise artists attempted to put their sound preferences on a pedestal. Both movements attracted feverish public attention. At the end of the 1930s, things took a turn for both kinds of noise elites, and not in a direction they would have wished.

To come to terms with this period of heightened movement, this chapter will discuss the publications of Russolo and Antheil, as well as those by artists and critics, most notably, the artist Piet Mondrian, who reflected on the music and musical instruments of both men.[1] Russolo's and Antheil's machine art certainly did not stand alone. Alexander Mosolow's *Steel Foundry* and Sergey Prokofiev's *Pas d'Acier,* both written in 1927, are two among many examples of machine-inspired musical compositions. After World War II, Edgar Varèse, Pierre Schaeffer and John Cage became the heroes of such music. This chapter concentrates on the work of Russolo and Antheil, however, because both made music out of machines, and because they were fiercely debated in contemporary art and music literature. Moreover, this chapter focuses on those aspects of their views that explain their love of particular characteristics of machines. Finally, I will analyze the predominantly European man–machine debate in music that arose in the wake of Russolo's, Antheil's, and Mondrian's publications.

THE SOUND OF MUSIC AND THE DYNAMICS OF MODERN LIFE

The question of how the noise of machinery became embedded in music is usually answered by referring to the Italian futurist movement. Textbooks on twentieth-century music define futurism as the "first clear manifestation" of "a major and enduring concern" for "the relationship between new music and modern technology" (Morgan 1991: 117). According to this commonplace view, the experimental nature of futurism inspired numerous composers "to capture the dynamics of the machine," "to extend the resources of sound," and to express the "love of mechanized urban life" (Watkins 1988: 243–249).

Fillipo Tommaso Marinetti and other futurists indeed wanted to wrench art from the past and follow the dynamics of modern times. According to them, new forms of technology, such as the telephone, gramophone, movie, train, automobile, zeppelin, and airplane had deeply influenced human experience, generating a complete renewal of human sensibility (Calvesi 1987: 11) Since "order, meditation and silence" had disappeared from life, art should likewise change (Marinetti, cited in Berghaus 1996: 17). *Life* itself, and most of all *movement* and *energy*, were the values futurists wished to express in their art.

These values, however, did not only reflect futurists' "bruitism," their partiality for the everyday phenomena of modern existence, but *also* their tendency to express a semispiritual "life force." As the painter Gino Severini noted in the Dutch art journal *The Style* (*De Stijl*), beauty was to be found in "universal movement," in "life." Therefore, futurist artists favored subjects such as motorcars or fast trains, because they enabled them to express the essence of reality. Such mechanical subjects, however, were not necessary. A trotting horse similarly contained universal movement (Severini 1919: 26).

Russolo's proposal to enrich musical compositions with noise was an extension of futurist ideas into the domain of music. For Russolo, everyday noise was "triumphant," and reigned "sovereign over the sensibility of men" (Russolo 1986/1916: 23). In a style resembling the auditory trope of sensational sound, Russolo lauded the sounds of the city: "Let us cross a large modern capital with our ears more sensitive than our eyes. We will delight in distinguishing the eddying of water, of air or gas in metal pipes, the muttering of motors that breathe and pulse with an indisputable animality; the throbbing of valves, the bustle of pistons, the shrieks of mechanical saws, the starting of trams on the tracks, the cracking of whips, the flapping of awnings and flags" (Russolo 1986/1916: 26). Machines, as Russolo put it, had created such a variety of sounds that traditional tones alone, because of their evenness and monotony, did not arouse enough emotion. To affect the modern listener required more than diatonism and chromatism. Hence, the timbres of noise, including those of modern war, should be added to the sound of traditional instruments. To phrase Russolo's views in a slightly different way, music should change since the modern experience of the sonic environment had changed. Music should broaden and perfect its scope.

Russolo's stress on enlarging and enriching rather than supplanting traditional sound explains the fact that his noise instruments or *intonarumori*[2] not only contained the acoustical phenomena of new forms of technology, but also those of nature. For instance, the "burster," produced a noise that sounded like "an early automobile engine," the sound of the "hummer" resembled that of "an electric motor or the dynamos of electric power plants," and the "howler" produced a sound similar to a siren. However, the "hisser" was an instrument that sounded like heavy rain, the "whistler" resembled the sound of wind and the "croaker" that of a frog (Brown 1986: 12). Russolo's wish to enrich rather than to replace traditional sound also makes understandable why he recommended the aesthetical quality of machinery and city noises by comparing them to the sounds of nature. Sounds like the roaring of a waterfall or the gurgling of a brook were traditionally considered to be pleasant (Russolo 1986/1916: 25).

At the same time, however, both Russolo and Marinetti defended bruitist experiments against the anticipated accusation of being a mere "imitation" of surrounding noise. "The . . . noise networks are not simple impressionistic reproductions of the life that surrounds us," Marinetti said, "but moving hypotheses of noise music. By a knowledgeable variation of the whole, the noises lose their episodic, accidental, and imitative character to achieve the abstract elements of art" (Marinetti, quoted in Brown 1986: 18). Russolo was similarly aware of the dangers threatening his music: "The Art of Noises would certainly not limit itself to an impressionistic and fragmentary reproduction of the noises of life. Thus, the ear must hear these noises mastered, servile, completely controlled, conquered and constrained to become elements of art." Only by mastering the infinite complexity of noise, "multiplying a hundredfold the rhythm of our life," by stirring the senses "with the unexpected, the mysterious, the unknown," one would truly be able to move the soul (Russolo 1986/1916: 86–87).

Russolo, who started his career as a painter and only later shifted his interests to music, was not the first futurist to propose new forms of music. Between 1910 and 1912, the Italian composer Balilla Pratella published several manifestos in which he turned against traditional forms, rhythms, and tonalities, and encouraged the use of microtones, intervals smaller than a semitone (Bartsch et al. 1986: 55). And even before Pratella's pamphlets, the composer Ferruccio Busoni's famous essay *Outline of a New Aesthetic of Music* had appeared, in 1907. Busoni wanted to do away with program music (music referring to extramusical reality), and instead advocated absolute music, free from the material limitations of traditional musical instruments. All possible steps had to be taken to create an "abstract sound, a technique without hindrance, an unlimited world of tones" (Busoni, quoted in Beaumont 1985: 90–91). Busoni's liberated world of tones included new divisions of the scale and divisions of a whole tone into microtones. He therefore mentioned the *Dynamophone,* a kind of electric organ created by the instrument maker Thaddeus Cahill that was able to produce Busoni's scales.

Much like Busoni, Russolo aimed to create music that employed all possible microtones—a subject that comprised two chapters his book, *The Art of Noises.* Just as Pratella had done earlier, Russolo used the word *enharmonicism* for his system of microtones. Conventionally, *enharmonicism* referred to enharmonic equal tones, like E sharp and F, or B sharp and C. On instruments with a tempered system, such as the piano, these tones are really equal. However, this is not the case with wind and string instruments. Russolo considered the tempered system to be inferior, and compared it to a "system of painting that abolishes all the infinite gradations of the seven colors"—knowing red, but no rose and scarlet lake (Russolo 1986/1916: 62). The howling of the wind produced enharmonic scales, in Russolo's terms scales of microtones, and the even richer world of machine noise was constantly enharmonic in the rising and falling of its pitch. Therefore, enharmonic instruments should be created to be capable of changing pitch by enharmonic gradations instead of diatonic or chromatic leaps in pitch.

MERE IMITATION VERSUS THE ESSENCE OF BEING

Russolo's innovations were debated extensively in the international art press. Some critics were discontent with the instruments, for instance because the differences between the varying noise groups were considered too small (Coeuroy 1920/1921). Others accused futurist composers of producing sheer cacophony, sensation rather than emotion. According to the music critic and composer Nicholas Gatty, they were breaking the laws of nature, since tonality, which is based on natural overtones, was grounded in the physical facts of sound. And how could variation be achieved when concordance and dissonance had both been abolished, and when there was no regular rhythm that gave melody its frame and music its form? "It is very difficult to believe," Gatty concluded, "that music composed of such materials, a series of negatives, can have any really emotional value, that it can affect us in any other way than by giving us nervous shocks of a sensational, physical order" (Gatty 1916: 11–12).

Some artists felt attracted by futurist theory, but were disappointed with the specific musical practices it inspired, including Edgar Varèse and George Antheil, and the painter Piet Mondrian. Marinetti's and Russolo's fears about being accused of just reproducing everyday noise proved to be right. Varèse had dreamt of new instruments like the ones that Russolo had created, but considered futurist music to be only a servile imitation of daily life (Varèse 1917). Mondrian and Antheil felt similarly. Remarkably, they accused the futurists of venerating everyday sounds of machines per se, whereas today's textbooks qualify the veneration of everyday sounds as precisely what machine art composers sought to achieve. To unravel this apparent paradox, one needs to understand that Mondrian and Antheil could accuse futurists of mere imitation because the artists involved in noise music deeply disagreed about which characteristics of machines had real artistic importance. Although they indeed shared an interest in technology and everyday noise, they had quite different views on what constituted the essential qualities of machines. For example, for Mondrian the idea of the essential qualities of machines were connected with his conception of nature and its place in music. Even more important for grasping the disagreements among these artists, however, were the differences between Mondrian's and Antheil's views on the immutable essence of life itself.

Mondrian's aesthetic ideal was *neoplasticism*. Neoplasticism advocated universality against individuality, activity instead of passivity, manliness over femininity, inner beyond outer, spirit at the expense of matter, and openness in place of closedness, even though Mondrian ultimately aimed, in line with his theosophist convictions, at a new culture that would integrate these oppositions into an harmonious whole. Mondrian felt that the plastic arts (painting and sculpture) had been predominantly individual: round, physical, bowed, closed—imitating nature. In contrast, new plasticism should be universal: abstract, internal, mathematical, and exact, as could for example be read from the straight lines in neo-plastic painting. Reality should not be denied, but instead be what was called "abstract-real," in which art expressed the deepest imaginable, pure, and immutable reality. Since neoplasticism involved all of the arts, music also needed to be renewed. For this reason, traditional instruments, whose full, vibrating, echoing tones and timbres were intended to imitate man's voice, reminding the listener of waves and curves, had to be replaced.

According to Mondrian, the futurists had done a good job of supplementing the timbres of traditional instruments with those of noise. However, he believed their experiments did not go far enough. Although they intended to make music more "objective" by making it more real, he contended that their music remained too close to nature and was not adequately abstract. For Mondrian, the names of their imitative instruments, the hummer, croaker, or whistler, proved this. The fact that the futurists' concerts in Paris, one of which Mondrian attended in 1921, presented the more traditional compositions of Russolo's brother Antonio rather than Luigi's networks of noises as expressed in *Awaking of a City* may have strengthened his opinion (von Maur 1981: 289). Mondrian also argued that the noise instruments were really no different from traditional instruments, since many of the futurists instruments could still perform diatonic and chromatic melodies and showed regression to the continuous fusion of natural sound. Nor were the bruitist instruments able to stop sound abruptly.

Mondrian's reproach of the futurists' music may sound odd given his interest in machine music. Yet for him, discontinuity was a prerequisite of neoplastic music: instruments should keep wavelength and vibration fixed and must be able to break off a tone suddenly. Instruments using electricity, magnetism, or instruments being automata were the best suited, because they excluded the human, individual touch, and enabled a perfect determination of sound. Even jazz was further ahead than the bruitists were. It was further removed from harmony and its percussion made a sudden intervention of intuition possible.

Mondrian stressed that futurist music had insufficiently freed itself from "repetition." He considered repetition a characteristic of both nature and the machine, although machines showed *accelerated* repetition and, therefore, enhanced objectivity, since Mondrian equated speed in music with straight lines in painting. Art, however, should *not* display natural repetition or its counterpart, "symmetry." On the contrary, art needed to be made up of nonsymmetrical relations: in painting, between color (red, yellow, and blue) and not-color (white, black, and gray), and, in music, between tone (sound) and not-tone (noise). Furthermore, successive tones should be different in volume and character. Therefore, Mondrian preferred the sound of traffic, which included irregular contrasts, to the repetitive rhythm of the carillon (Mondriaan 1921a; Mondriaan 1921b; Mondriaan 1917; von Maur 1985; von Maur 1981). At least the rhythm should be tight, such as in the *Stijlproeven* (exercises in composition) of his friend and composer Jacob van Domselaer, with whom Mondrian exchanged ideas about music (Bois 1994).

So unlike Russolo, Mondrian did not stress the enharmonic quality of machines and thus their ability to produce *continuity,* but their capacity to *fix* and *break off* tones, in order to create a tension between tone (sound) and non-tone (noise). Whereas the enharmonicism of Russolo had its counterpart in the infinite gradations of the color spectrum, the sound of Mondrian had its counterpart in primary colors. Moreover, Russolo saw no problem in associating his noise networks with nature, since nature, just like technology, was part of modern life, and since comparisons with nature could justify his experiments, making the latter seem less strange. Mondrian, on the contrary, did everything to disassociate art from nature because for him art should be abstract-real. At the same time, however, the machine, *as such,* did *not* embody the new culture that he strived for. Abstract reality could only be expressed through machines capable of creating a rhythm both fast and irregular, in contrasting relations.

George Antheil, a member of the art movement *De Stijl,* similarly kept aloof of bruitism. He clearly resembled Mondrian in his critical stance toward futurism. And as van Dijk (1996) has asserted, both Mondrian and Antheil had an interest in the ability of machines to realize a tight, precise rhythm, and a fascination with movement and time, Mondrian's counterpart of space in painting.

Their ideas on music were not completely in line, however. They may have had similar intentions with respect to precision, that is, to produce abstract,

147

streamlined, and tense music. However, they had rather different ideas about
how to achieve it. According to Mondrian, traditional music attempted to create
contrasts by repetition and rests. Since the emptiness in the rests was immediately filled with the listeners' individuality, Mondrian believed that new music
ought to consist of a constant expression by rapidly alternating tone (sound) and
nontone (noise) (Mondriaan 1921b: 133).

Antheil in contrast made silences and repetition (of rhythm and by creating loops in compositions) key elements of his *Ballet Mécanique*. Time itself, he
said, had to work like music. For him, the most important feature of futurist
music was not the microtone, but time. Antheil considered his *Ballet Mécanique*
as an example of a new fourth dimension of music, the first "time-form." For
him, time was the sole canvas of music and not a by-product of tonality and
tone. Time was inflexible, rigid, and beautiful, the very stuff out of which life
was made. According to Antheil, never a modest man, *Ballet Mécanique* was the
first work to ever have been composed out of and for machines, neither tonal
nor atonal, just made of time and sound, without the traditional contrasts in
piano and forte (Antheil 1924; Antheil 1925a, 1925b).

In his autobiography *Bad Boy of Music* (1945) Antheil explained that he had
chosen the title *Ballet Méchanique* since these words were "brutal, contemporary,
hard-boiled, symbolic of spiritual exhaustion, the superathletic, non-sentimental
period commencing 'The Long Armistice.'" But he had certainly not meant
the piece as "a mundane piece of machinery." Although he loved machines for
their beauty, he did not want to copy "a machine directly down into music"
(Antheil 1990/1945: 139–140). Nor did he want to use machines like the ones
the Italian futurists employed, since "these had no mathematical dimension at
all, nor claimed space, but just improvised noise that imitated motorcars, airplanes etcetera, which is ridiculous and had nothing to do with music." (Antheil
1924: 101)

Antheil's first version of *Ballet Mécanique* had included a score for sixteen
player pianos, which according to Whitesitt "run electrically from a common
control" (Whitesitt 1983: 106). The complex, mechanical ballet was meant to

accompany a film, for which Fernand Léger would write the screenplay. But in the end, Antheil's score was twice as long as the movie, and profound difficulties arose in their attempts to synchronize the sixteen player-pianos, so neither the collaboration with Léger (although the latter did make his film), nor the player-piano version of the musical composition was ever realized. Instead, Antheil decided to orchestrate his ballet, using—among traditional pianos and percussion—one player piano, airplane propellers, sirens, and electric bells. After the debacle of the composition's Carnegie Hall premiere, Antheil—still only in his twenties—changed his style to neoclassicism just as Stravinsky had done. His reputation constantly hovered between that of a "publicity-seeking rogue, charlatan and insincerist" (Thompson 1931: 20) and a very talented, but not yet important composer (Copland 1925; Petit 1925).

Notwithstanding such ambivalences in public esteem for his compositions, Antheil *did* contribute to the use of machinery noise in music. Antheil considered other characteristics of machines significant to music than Mondrian did, while Mondrian felt attracted by different aspects of machines than Russolo. Mondrian believed the repetitive sound of machines was dangerous, reminding him of nature, and similarly risky was silence. Instead, machines should enable a controlled *irregularity.* To Antheil, however, both silence and *repetition,* which machines could produce so wonderfully, were paramount to time-music. Moreover, neither Antheil nor Mondrian showed any particular interest in microtones, which Russolo considered a fascinating quality of both nature and machines.

These divergent views on the principal qualities of machines make understandable why Mondrian and Antheil both accused the futurists of imitation, whereas futurists *themselves* were convinced of the abstraction of their noise networks. Since Mondrian and Antheil had chosen different characteristics of the machine for artistic expression than Russolo had, they heard only the noise of machines in Russolo's musical instruments and compositions. And although Russolo, Mondrian, and Antheil unanimously stressed the capacity of machines to be *precise,* their beloved precision served different ambitions with respect to abstraction. Abstraction was to be found in the infinite, enharmonic complexity of noise (Russolo), the fast and irregular alternation of tone and nontone (Mondrian), or the mathematical time-form of repetition and silence (Antheil).

Moreover, these abstractions were far from noncommittal, since such music had to make audible divergent conceptions of the fundamental essence of life: a maximized or multiplied reality (Russolo), an immutable universality (Mondrian), or rigid, inflexible time—the very stuff out of which Antheil believed life was made. These all-embracing, basic, and constant values again explain why Mondrian and Antheil failed to appreciate the musical works of the futurists. Their machines had to express other ontologies than those of Russolo.

It has become commonplace to define the work of Russolo, Antheil, and Mondrian as thoroughly modern because they seemed to break with all that had been sacred in the romantic music of the nineteenth century. Professor of music Richard Taruskin, however, argues that the search for maximum expression in early-twentieth-century avant-garde music should be seen as a form of "late, late Romanticism." Only with the introduction of neoclasscism did modern music, with its stress on objectivity, really begin (de Bruijn 2005: 30). Reinterpreted in this light, the noise artists' aim of allowing the machine embody the essence of life was a romantic rather than a modern ambition. Even the significance that these and other proponents of machine music attributed to mechanical precision, an issue further elaborated on below, expressed romantic ideal. The precision they expected of machines gave voice to the composers' intentions, thereby reiterating "the Romantic enthronement of the autocratic and infallible composer-creator" (Taruskin 1995: 167). And this, in turn, linked the noise artists' endeavors to the elitist Schopenhauer-attitude that had dominated early-twentieth-century noise abatement debates.

Indeed, avant-garde artists and noise abaters encountered a strikingly similar gap between receiving considerable press attention and achieving only a niche-confined success. Antheil had his five minutes of fame, but never acquired a stable reputation for being a good composer. Between 1913 and 1927, Russolo's concerts were a success, with some thirty thousand people attending a series of performances in London (Bartsch 1986: 21). However, his instruments turned out to be a commercial failure. For a while Russolo made money with *intonarumori* by using them as the accompaniment for silent movies. But the rise of talkies quickly put an end to this. Sadly, only two pages from the score of *Awakening of a City* and none of the *intonarumori* have survived. Some of the intonarumori were probably cannibalized by Russolo himself for new uses, but others were lost during World War II. Russolo eventually took up painting again and became involved in mysticism and theosophy. He died in 1945, an ascetic recluse, having turned his back on the noisy world that had once inspired him.

THE MAN-MACHINE DEBATE IN MUSIC—CREATIVITY AND CONTROL

Although Russolo, Antheil, and Mondrian differed sharply with respect to the ontologies they wanted their machines to express, they were alike in wanting their machines to be very precise so that the sounds they produced could be controlled. In fact, *control* was one of the most significant issues discussed in the literature on machines and music in the 1920s and 1930s. In 1924, the composer and critic Hans Stuckenschmidt prophesized that the traditional orchestra would be done away with in less than fifty years and replaced by new "mechanical" instruments, or automata. Many classical musicians and critics considered Stuckenschmidt's stark prediction blasphemous. It provoked fervent discussion in musical circles in Europe, and musical journals such as *Der Musikblätter des Anbruch* (The Sheet Music of Dawn) and *Der Auftakt* (Upbeat), felt the subject was important enough to devote a special issue to it. The famous avant-garde music festival held at Donaueschingen every year dedicated a large part of its 1926 program to showcasing mechanical music, inviting several composers to compose a score for mechanical instruments and inventors to demonstrate new instruments.

How did Stuckenschmidt support his shocking and widely published forecasts? First, he argued that classical music concerts were so poorly attended that symphonic orchestras had been forced to depend too heavily on public and private funding for their financing. This, he believed, would lead to their demise in the long run. Second, orchestral scores had become increasingly complex, which was driving musicians into despair. Some orchestral works had already become unperformable, because the intentions of composers had exceeded human's capacities of interpretation. Third, individualism had become outdated. The new age asked for a more collective attitude. Consequently, new music required "objectivity." Once the limitations of instruments and the mistakes of performers had been removed—out of tune violins, the clattering valves of wind instruments, hesitant horn-players, out of breath oboists—composers could employ new modes of musical expression. Examples of these innovations were the use of mathematically perfect intervals, such as those demanded by the modern twelve-tone system, quarter- and other microtones, complex rythms, and the simultaneously single-handed playing of different timbres. Since it was still impossible to prescribe a composition's performance in detail, except for tempo and dynamics, it followed that composers would continue to search for a definite consolidation of performances by new means of music notation.

Machines met all these new requirements. "The machine," Stuckenschmidt claimed, "has no limitations. Its strength and speed are practically limitless, its performance is of unfailing precision and uniform objectivity" (Stuckenschmidt 1927: 9). New mechanical instruments had to combine the characteristics of the orchestrion with that of the player piano, since music stamped directly on a paper roll produced music with ideal precision, without a performer's intermediate individual interpretation. Even the phonograph and the gramophone had a bright future as musical instruments, since they could create any tone-color or rhythm, even ones that were non-existent in modern orchestras. Just as Busoni had done previously, Stuckenschmidt also mentioned Thaddeus Cahill's *Dynamophon*. The *Dynamophone* was able to produce all conceivable timbres in any desired pitch, tempo, and arrangement by means of electrical vibrations.

New instruments could thus not only solve orchestras' growing financial problems (since mechanical instruments were cheaper) and overworked musicians (since machines could produce the most complex music reliably), but the problems of age as well. The era strived for objectivity, but had been hindered by imperfect notation-systems and instruments. Stuckenschmidt certainly did not expect to enter the world of music unopposed. His adversaries might say that human interpreters remained necessary: there could be no phonograph without a performer. Against this argument, Stuckenschmidt contended that only a single performance of any musical score was necessary for making a master recording, and that one could even produce a record graphically, like an etching, without involving a human interpreter. His opponents might feel that music produced in such a manner was like dead mechanics and soulless automats. Stuckenschmidt, however, stressed that machines could perform as excellently as human interpreters. Had not Stravinsky arranged *Les Noces* and *Sacre du Printemps* for the mechanical Pleyela-Piano, and George Antheil conceived his *Ballet Mécanique* for mechanical interpretation? Musicians, Stuckenschmidt admitted, would indeed be deprived of their daily bread, but not for another fifty years (Stuckenschmidt 1924; Stuckenschmidt 1926a; Stuckenschmidt 1926b; Stuckenschmidt 1926c; Stuckenschmidt 1926d; Stuckenschmidt 1927).

In short, Stuckenschmidt underlined the machines' potential to cope with *complexity* in musical compositions and to maintain *precision,* a quality that had been of similar significance to Mondrian, Antheil, and Russolo. Yet Stuckenschmidt also admired the ability of machines to create objectivity, an objectivity that perfectly matched the collective, modern attitude of his age. Many of Stuckenschmidt's progressive colleagues agreed with the need for objectivity. As they saw it, machine music followed from the trend to compose a-sentimental, objective, motorial music (Holl 1926; Steinhard 1926; von Strassburg 1926; Toch 1926). Others saw a particular potential of mechanical music when composed for radio and gramophone—technologies that needed rigorous, linear, and rhythmic music (Toch 1926; Schünemann 1931).

Still, not all was modern behind the search for machine-music. Machines could produce a controlled objectivity, and the control they enabled one to have must even bypass the performer's interpretation. Yet the ambition for control, as we have seen, articulated a thoroughly romantic notion, since it was all about etching the great minds of composers into the machine. While proponents of mechanical music wanted to protect the voice of composers against that of performers, opponents to mechanical music sought to protect the cultivated art of performing against the crude taste and boring music of the masses. The issue at stake was who was or was not allowed to control the sound of music?

The introduction of the player piano in the United States at the turn of the twentieth century had been accompanied by fairly critical comments. Player pianos, such as the *Pianola* and the *Welte-Mignon,* produced music mechanically through a set of instructions stored on a perforated music roll. Although the player piano's proponents in the United States, which included piano manufacturers and publishers of sheet music, thought that it would lead to "an almost universal music education," many music teachers, musicians and composers opposed it (Roell 1989: 39). Opponents to it contended that one could copy sound, but not interpretation, and that mechanical instruments reduced the expression of music to mathematical systems. For this reason, they believed that mechanized music diminished the ideal of beauty by "producing the same after same, with no variation, no soul, no joy, no passion," and that the introduction of the player piano would lead to the disappearance of amateur players (Roell 1989: 54). One commentator feared that those belonging to the "lower level" of society would copy the "cultivated class" to advance themselves (Roell 1989: 57). Others feared that music would "lose its distinctiveness, its uniqueness as an experience." After "several months of mechanical music," a critic said, a listener would miss the "hesitating sounds which once charmed him, that human touch which said something to him although imperfectly" (Roell 1989: 58).

Opposition to the player piano and other mechanical instruments was equally vehement in the European music world. Taking a stance that was similar to that of their American counterparts, European critics held up the beauty of imperfection as an argument against mechanical instruments, and characterized mechanical music as dull, tiresome, cool, and frozen (Jemnitz 1926; Krenek 1927; Preussner 1926; Felber 1926). Art could "consist only in communication ... from soul to soul, a communication that the machine is, and ever will be incapable of creating" (d'Indy, cited by Coeuroy 1929: 265).

Different views about the significance of the musician's role in interpreting a piece of music led to heated debates. According to the composer Ernst

Krenek, the inhumanly "clarity, exactness and preciseness" of mechanical music had been fully acknowledged as compositional principle. He hesitated about its value, however. He felt that mechanical music left no room for the "primitive drive of musical man" for active participation in musical performance (Krenek 1927: 380). A jazz band performing complex polyrhythms with mechanical exactitude created artistic excitement. But no such thing happened when a machine was used, for a machine was simply expected to be precise. The Dutch composer Willem Pijper observed that the speed, power, and polyphonics of the player piano meant "less than nothing [compared] to the mystery that was the living centre of all art" (Pijper 1927: 33). Another Dutch musicologist and critic, Herman Rutters, noted that the mental process of overcoming technical difficulties in playing a musical instrument was crucial to ones' artistic development. The true artist was aware of this, whereas the dilettante was not (Rutters 1913).

Some mechanical instruments, such as the Welte-Mignon, indeed enhanced a composer's control over the interpretation of music. Whereas the music rolls for the player piano were produced by recording interpretations of famous performers, compositions for the Welte-Mignon could be punched directly onto the music roll (Heinsheimer 1926; Cowell 1930–1931). This was the innovation that Stuckenschmidt had been so fond of. With the help of machines, a composer could exclude the mistakes that performers made and consolidate his interpretation. Not all composers valued this approach. Arnold Schoenberg, for instance, argued that the composer's own interpretation should never be the final one (Schoenberg, cited by Äusserungen 1926). One contributor to the man–machine debate in music even denied that the composer actually achieved this sort of control over a performance, since the player piano remained sensitive to the particular circumstances in which it was played, responding, for example, to changes in temperature (Haahs 1927).

In *Mensch und Machine* (Man and Machine) and *Der Entgötterung der Musik* (The Disenchantment of Music) Adolf Weissmann formulated one of the most comprehensive attacks to be published on mechanical music. He lamented that both the machine age and psychoanalysis, another sign of the times, had destroyed romanticism. The machine, with the aeroplane as its key icon, had reduced the secrecy that had been key to romantic art. Bringing everything within reach, the machine facilitated a process of equalization. Psychoanalysis did exactly the same thing. It transformed secret dreams into unimaginative, down-to-earth confessions, endangering creativity in the process. Psychoanalysis also aimed to free individuals from inhibitions that hampered the expression of their musical capacities, a misguided attempt, as Weismann put it, to democratize "the spirit" (Weissmann 1927). Even worse, performances without human performers would never touch the audience (Weissmann 1928). Others identified similar problems with stage performances by mechanical instruments. After witnessing a concert of a Welte-Mignon at Donaueschingen, chief editor of *Der Auftakt,* Erich Steinhard noted that the public hesitated about what to do after the performance had ended: "Should one applaud? For nobody is there. It is only a machine" (Steinhard 1926: 184).

In mounting their opposition to mechanical instruments, such critics made their views explicit as to what real art and interesting music should be about. The need for a romantic passion and creativity and the fear of music's democratization were the key issues at stake in their arguments against these new instruments. Performers and music teachers probably also feared for their livelihoods. Innovations in art often threatened artists and performers who had mastered their skills through time-consuming learning processes, as Howard Becker has noted in *Art Worlds* (1982: 306). The paradox, however, is that the proponents of mechanical instruments, at least many of the composers among them, were no less romantic in their wish to control sound and thus defend the genius of a composer against the intrusions of a performer.

The most ironic comment on the use of machines in music came from the American music critic Irving Weil. He did not like the music of Antheil and his companions. Yet Antheil was perhaps ahead of his time. In the future, man's sense of hearing would become muffled if the noise of everyday life continued to get louder at the current rate. In half a century or so, man would be "literally pig-eared," if not deaf, which would require the use of noisy musical instruments (Weil 1927). This was a dreadful outlook for Weil, but Russolo, Antheil, Mondrian, and Stuckenschmidt would have welcomed it, at least with respect to the instruments that would have to be used.

———

Conclusions

Textbooks about twentieth-century music rightly stress that many of the artists promoting the use of machines and mechanical sound in music were inspired by futurists in their fascination for the noise and dynamics of mechanized, urban life. However, textbooks omit the fact that several of these "machine artists" *also* followed futurists in their effort to express a semi-spiritual life force. And since these artists disagreed about *which* machine characteristics best expressed the deeper essence of life's reality, they disapproved of each other's use of machines in music. For this reason Mondrian and Antheil accused their predecessor Russolo of merely imitating environmental noise, even though they had clearly been inspired by futurists' work themselves.

To make life's essentials audible, artists stressed various and often opposing characteristics of machines. *Continuity* (the production of microtones) and *discontinuity* (the sudden break off of tones), the ability to create controlled *irregularity* and controlled *repetition,* as well as *precise complexity* were all considered to be the most significant capacities of machines. Russolo, Mondrian, Antheil, and Stuckenschmidt had different things in mind when talking about the essence of life, or in the case of Stuckenschmidt, the basic feature of the age. Russolo sought to maximize or multiply reality, Mondrian tried to capture an immutable universality, Antheil tried to bring time to the fore, and Stuckenschmidt strived for objectivity. Thus, the machine in music did not only function as a means of glorifying the speed of modern times, but was simultaneously felt to vocalize reality's fundamentals, the real essence of life. Because *both* of these issues were at stake, the machine *itself* did not embody the new culture, instead specific *characteristics* of the machine did.

For each of the four artists, one of these characteristics was the ability of machines to create complex music in a precise manner. In general, clarity and exactitude had become increasingly important as compositional principles. To Stuckenschmidt, such technical mastery even meant that the composer could get rid of the imprecise performer, which implied a highly elitist control of making sound, thus sharing more with the refined mind-tradition in the noise abatement movement than one would expect at first sight. Although the exclusion of the performer was controversial, many postwar composers valued control over interpretation, which created the should-one-applaud problems that electronic music on stage has to cope with to this day.

Outside the concert hall, however, music making *did* become democratized. Although the piano player did not survive as a musical instrument, the phonograph and the gramophone became culturally embedded as the mechanical musical instruments of the masses. The notion of control over musical sound—who was and who wasn't allowed to make music—would become a topic of vital significance in the debate about the noise made by neighbors. The entitlement to making music transformed into a kind of inalienable right of all mankind and would increasingly hamper the abatement of domestic noise.

A WALL OF SOUND:
THE GRAMOPHONE, THE RADIO, AND THE NOISE OF NEIGHBORS

THE NOISE OF NEIGHBORS BEYOND LAW

"There is no doubt," the British physicist A.H. Davis declared in 1937, "that noise is one of the features of uncontrolled development in a mechanical age— an undesired by-product of the machines which are increasingly employed for industrial and even domestic operations. . . . In some residential quarters, in such quiet as the evening possesses, . . . some owners of gramophones and radio-sets seem oblivious of the fact that their instruments are so grotesquely over-loud that speech items are audible and intelligible in a muffled sort of way for several yards up and down the street" (Davis 1937: 1–2).[1] It was indeed the "mechanical age" that set the stage for the gramophone, the radio, hi-fi stereo sets, and all kinds of electrical devices that radically changed the nature of the sounds produced at home by residents as well as by their neighbors. These sounds triggered new public debates on the ways in which auditory privacy should be or could be secured, particularly in town and city life. In such debates, a wide range of issues related to community culture, private space, and scientific and technological developments converged.

Science and technology have played crucial roles in regulating and resolving the potential conflicts and problems resulting from urban overcrowding and lack of space. In late-nineteenth-century Paris, for instance, the science of the "germ" helped legitimize formerly inconceivable interventions in private homes that were aimed at preventing epidemics and securing public health (Aisenberg 1999). In twentieth-century urban traffic control, the responsibility for regulating the enormous flow of citizens through a limited number of streets has been largely delegated to machines such as traffic lights. Moreover, all kinds of science-based criteria, like blood alcohol level and maximum speeds, have meanwhile objectified norms of behavior in cities and elsewhere. In a more general sense, this kind of "trust in numbers" within a "culture of control"—which in some urban contexts has reached sheer perfection—has been a major concern in the history of science and technology (Porter 1995; Levin 2000).

This chapter, however, focuses on a conflict in which our reliance on objective scientific standards failed to be productive: the case of domestic, or neighborly, noise. In this social context our confidence in numbers and technology has *not* been rewarded. Even though the decibel eventually became a major factor in measuring and controlling traffic noise, noise level–based legislation aimed at regulating disturbance from neighbors has never been realized in any comprehensive manner, except for the soundproofing of housing. Today, we have quantified norms for the sound insulation of walls, but not for regulating the level of noise produced by neighbors. This chapter aims to explain why this was the case. In fact, the chapter reveals some developments that have had lasting effects on Western societies' handling of noise problems.

More specifically, this chapter aims to understand why Dutch attempts to solve controversies over noise from neighbors on the basis of scientific or technology-related standards failed. In the mid-1970s, the Dutch government expressed its frustration with the fact that it was impossible to capture problems associated with individual cases of neighborly noise "inside a legislative system" (The Soft 1979). In the mid-1980s, local police were required to intervene about 70,000 times, primarily in response to neighborly noise produced by modern technical devices (van Rossum 1988: 29–30). Today, nearly one out of three persons in the Netherlands claims to be bothered by neighborly noise, notably by the sounds from the radio, stereo, or television (Statistisch Jaarboek 1994: 91;

Bestrijding 1998: 1, 3). The country's Noise Abatement Act, however, does not encompass the noisy behavior of neighbors. In 1979, when the act went into effect, the government chose instead to address the issue of neighborly noise by subsidizing a huge public information campaign that they called "Let's be gentle with each other."

This chapter views this policy as the result of three long-term developments all of which began in the first half of the twentieth century: changing class relations, the shifting status of the subjectivity of sound perception, and the recognition of the right to make noise at home. All three trends played a role in preventing neighborly noise from being regulated by technology and standardized scientific measurement. Before elaborating on these trends, I will provide a context for the rise of the public debate on neighborly noise by focusing on the changing sound environment and noise awareness in early-twentieth-century residential areas.

A WORLD OF SOUND, A WORLD OF PARADOX

"Ears and hearing to-day takes pride of place of the five senses, because all the latest inventions are in the world of 'sound'—so that wireless (television), talkies, telephone, traffic makes hearing at its best an absolute necessity" (Noise 1935: 65). In the first half of the twentieth century, the Western world of sound was a world of paradox. Talkies, gramophones, radios, and telephones offered people's ears an exciting new array of sounds. Yet the ways in which those inventions were used could deafen those very same ears—at least this is what people believed.

This paradox itself may be one of the reasons why noise from neighbors became an issue of concern for Dutch citizens in the years after 1910. For instance, residents wanted to be able to hear the person with whom they were speaking on the telephone, but sometimes they had trouble doing so because of the noise coming from other devices that their neighbors were using at that very moment. Or, when they wanted to listen to the radio, nearby electrical machinery often interfered with the reception (Verberne 1993: 44, 48, 52).

161

Loud radio sets or people shouting into their telephones might even interrupt residents' conversations.[2]

Another change that was relevant for the genesis of neighborly noise as a public problem was the rise of noise abatement across Western Europe and North America in general. In the same way that the social elite had attacked traffic noise, citizens in the West sought to abate the noise of mechanical instruments, such as the gramophone and the radio: as unwanted sound produced by thoughtless individuals who were in dire need of public education. For instance, in 1930 the Noise Abatement Commission of New York proudly claimed in its first report that it had successfully asked "the radio stations of New York City to aid us in a campaign to educate radio listeners in noise etiquette." Each night at 10:30, stations throughout the city asked listeners to turn down their loudspeakers "as an act of good sportsmanship" (Brown et al. 1930: 55). A noise survey initiated by the Commission, which generated more than 11,000 responses from city residents, showed that radio ranked third on the list of the most reported sources of annoyance. Only the noises produced by traffic and transportation were viewed as worse (Brown et al. 1930: 27). As one of the press responses to the commission's report asserted, something had "to be done about the radio corsair who forgets the health, the privacy and the sanity of all the rest of the world in his own blind and crazy devotion to the metallic blatancies of the horn." Exacerbating the problem was the growing practice of listening to the radio while engaged in other activities (rather than just sitting down to listen), a practice that, as Susan Douglas has shown, was already quite common in the United States by the early 1930s (Douglas 1999: 84). As we have seen in chapter 3, this habit had been strengthened by the introduction of music on the shop floor. The appearance of these new listening practices meant more and longer periods of radio sounds. Neighbors, antinoise campaigns pointed out, were frequently forced to listen in, and this did not necessarily please them.

That many commentators saw towns and cities as the main loci of such nuisances is understandable. In densely populated areas, the sound of mechanical devices could be heard all over. Yet, as Theodor Lessing had suggested as early as

1908, it was also possible to run into the sound of a gramophone in a remote valley in the Alps. Therefore, it was not exclusively the spread of gramophones and radios that caused the uproar over noise from neighbors. Nor was the loudness of mechanical devices the sole reason for complaints. As Emily Thompson has claimed, it was the newness of particular sounds, even more than their loudness, that prompted citizens to action (Thompson 2002).

Several other factors that were characteristic of urban areas likewise contributed to the annoyance new technologies could create. First, as Alain Corbin has made clear for France, late-nineteenth-century cities were facing a remarkably de-standardization of the rhythm in everyday life. From the 1860s onward, city dwellers began complaining about the ringing of bells in the early morning, because of "a greater determination *to lay claim to one's morning sleep.*" Developments such as street lighting, "the circulation of elites within the town," and "the novel presence of women in public space together produced a gradual modification in nocturnal behavior. An enhanced desire for individual liberty prompted challenges to standardized rhythms" (Corbin 1999/1994: 302). The desire for individual liberty led to protests against bell ringing and also helps to explain the increasing public commotion over neighborly noise.

Second, the general focus on the significance of hygiene for people's health brought not only fresh air into homes, but noise as well. As Dan McKenzie suggested in *The City of Din,* people had been confronted with "a triumphant crusade in favor of the open window." Tragically, however, "in noisy streets sound cannot be kept out if the windows are open" (McKenzie 1916: 109–110). Moreover, the same quest for hygiene had led to new sanitary and heating facilities. The newly installed plumbing and heating systems, however, transmitted sound to every corner of an apartment building. The desire for better hygiene had also led to changes to the interiors of rooms such as smooth surfaces and floors, simple furniture, and less decoration—all of which had surprisingly negative acoustic consequences. In the words of Frederic Charles Bartlett, professor of experimental psychology at Cambridge, hygiene was "won at the cost of needless mental irritation" (Bartlett 1934: 76–77; Koelma 1929: 336).

Furthermore, many experts held that the employment of reinforced concrete had worsened the acoustic quality of private houses. "I regret to say," McKenzie wrote, "that many modern houses, particularly those which are made into tenements or flats, are not nearly so well segregated from each other's noises as the older houses are" (McKenzie 1916: 109). Dutch engineers claimed that in reinforced concrete and iron framework buildings, the vibration of mechanical moving power could easily manifest itself (Schoemaker 1929: 275; Beton 1929). This was due not to the use of concrete per se, but to the fact that buildings had become monolithic structures based on a single construction principle, and no longer consisted of the variety of building materials found in older structures (Zwikker 1936).

Because of these social and technical changes, neighbors began to play a more audible role in the life of town and city dwellers. Dutch citizens who felt too bothered by their neighbors and wanted to do something about it found some legal recourse in local ordinances as well as in the Dutch Penal Code. Since 1881, the penal code had included two sections on disturbance. The most important one was section 431, which stipulated that it was illegal to disturb people's sleep at night, whether by raising tumult that threatened the peace of one's immediate neighbors or by disturbances that broke the quiet of a complete neighborhood. According to the section's explanatory memorandum, the regulation was aimed not at noise caused by professionals who had to work at night or at people who attended balls and concerts, but rather at noise caused by people engaged in mischief. More specifically, the law sought to prevent yelling, cursing, or other rowdy behavior from intoxicated or overly excited people (van Dam 1888: 40–41).[3]

The rise of the gramophone and the radio in the first decades of the twentieth century created new potential noise problems between neighbors, one that the existing legal framework in the Netherlands had trouble addressing effectively. In my discussion below, I focus on the debates about this new technological noise that took place in the municipal councils of eight Dutch towns and cities.[4] It will become clear that it was far from easy to find common ground.

DO NOT SILENCE THE INSTRUMENTS OF COMMON MEN

The gramophone came up for discussion in Dutch local politics as early as 1913 in Rotterdam. This port city had recently witnessed spectacular expansion: between 1869 and 1913 its population quadrupled (Bank and van Buuren 2000: 124–126). According to the Rotterdam Committee of Penal Regulations, the time had come to ban the use of what they referred to as "mechanical musical instruments" in specific circumstances. These circumstances included the use of loud gramophones on or near "homes, buildings, halls, or structures, balconies, porches, and open appurtenances that form part of them or belong to them, as well as on vessels lying by or near public quays." Also included was the use of those devices *inside* homes, buildings, and so forth, in cases where the sound reached the outdoors through open doors, windows, or other openings.[5] The committee claimed that so many people made such "immoderate'" use of the gramophone that neighbors who were forced to listen in "for hours" could barely stand the nuisance.[6] Yet, the committee also stressed that authorities should be hesitant about interfering with people's activities at home. Hundreds of thousands of people lived in close proximity to each other, and they should impose a certain measure of self-restraint. When residents failed to do so, however, authorities needed to issue rules in order to ensure an orderly and undisturbed community life.

The committee's proposal caused a heated debate in Rotterdam's city council. Leftist council members and a few others strongly opposed the new ordinance. Their main argument was that the gramophone—a fairly inexpensive source of music—had become "the musical instrument" of the lower classes, and that an exclusive ban on gramophone-induced noise would therefore hurt them in particular. One did not hear gramophones, a critic claimed, in the city's upper class districts. To underscore his point, he told an anecdote about a man from one of the poorer districts who had won the lottery. "The first thing he did ... was to buy a gramophone. Now the neighbors enjoy it so much [as well], that they put some money aside each week in order to be able to buy a new record for it."[7] As another opponent of the proposed regulation put it, rather than driving a wedge between neighbors, the gramophone brought them together.[8]

According to one of the more conservative council members, families also benefited: "I have been frequently pleased by the fact that such an inexpensive device can keep a family together, whereas its members might otherwise each do their own thing and probably be tempted to seek entertainment that is less appropriate, albeit less of a nuisance to neighbors." The same member pointed out that although "the incessant audibility of gramophone sounds can be a great nuisance to many, ... one can also be desperately bothered by lady singers, trombonists, pianists, cellists and any other 'ists.'"[9] In a similar vein, leftist council members argued that it would be unfair to ban the loud noise produced by "phonographs," while allowing the sound of "mal-treated" pianos and house organs, as well as the "miserable lamentations" of concertinas, to go unregulated.[10] Another council member also concluded that one should question the fairness of excluding "only one of the musical devices, almost invariably the gramophone, [while leaving] the generally more substantial nuisance of other instruments unregulated."[11]

Most Rotterdam city council members agreed, however, that the "hobby" of those who put gramophones in front of open windows was a "debauchery." They insisted that there was an essential difference between music made by musicians and music generated by a device. "After an hour's practice, flutists, oboists, and so on ... feel a need to do something else, like smoking a cigar, talking to someone or devoting themselves to some other study. They get tired and stop playing and therefore they cause less hindrance, nuisance, harassment, and irritation."[12] After a lengthy discussion, the city council passed the proposed ordinance with a vote of twenty-four to thirteen, giving the local government the power to interfere in situations in which people caused nuisances by using loud gramophones.[13] Subsequently, on the instruction of the provincial government, the city council had to put in the important restriction that police officers had the right to enter a house against the will of the inhabitant only when accompanied by high officials.[14] This restriction, though, did not change the basic concept of the ordinance.

The first attempt by local authorities in the Netherlands, then, to solve the problem of gramophone noise—by differentiating between music produced by indefatigable machines, thus *technology,* and music produced by humans with limited energy—was accepted. The difficulty council members experienced in weighing the various pros and cons, however, highlights both the legal limits they had to reckon with and the resistance of those who considered such measures an elitist form of noise abatement. The conservative council members won out, but the simple reference to "mischief," as in section 431 of the Dutch Penal Code, was not enough anymore. Significantly, the council members representing the working class conceptually carved out, so to speak, the right of workers to a "sound culture" of their own. They did so by referring to the low price of the gramophone, by characterizing it as a "musical instrument for the masses," and by creatively redefining "making noise" into "sharing music." In addition, others reframed the "menace" of such sound from an issue about noise that disturbed the peace of a neighborhood to an issue about the danger of depriving families of a device that could secure them from disintegration or engaging in socially less desirable activities.

In the 1920s and 1930s, the city councils of The Hague, Leiden, Breda, Utrecht, Amsterdam, Maastricht, and Groningen followed Rotterdam in taking measures against noise generated by musical devices, and extended them to include the radio.[15] In general, the more progressive council members, especially communists, social democrats, and left-leaning liberals, declared themselves to be against the new ordinances, whereas the more conservative ones usually supported such measures.[16] The latter wanted to do something against "the maniac or brute who dominates everything in his neighborhood, shoos away the birds from the trees, yaps away the good and modestly recited music and imposes his taste upon others, prevents children from sleeping with open windows and disturbs the ill." They sought to ban the "stupid, machine-like" playing of records and radio for hours and hours without listening.[17]

A number of socialist council members stressed that it was in the interest of working-class children to get enough sleep and that working-class adults should have the opportunity to unfold their talents during the evening, which meant that they should not be annoyed by radio noise.[18] Yet, most of the left-oriented council members stressed again and again that it was not reasonable to limit the pleasures of the gramophone and the radio while allowing piano and organ players to annoy their neighbors endlessly. "It strikes me," a communist Amsterdam council member asserted in 1929, "that every time a certain commodity becomes available to working class people, effectively ending its monopoly in the hands of the bourgeoisie, measures are taken to limit the use of that new commodity." He added that workers who worked in shifts might not get off of work until late in the evening. Why should they not be allowed to enjoy the radio at night? He and other council members stressed that in working class areas, radio noise was not considered a problem at all.[19] "When I go home," the Amsterdam communist said, "I always enjoy being accompanied by music from the same radio station coming from every house of the street."[20]

The argument that the gramophone and the radio were typical working class instruments was not merely a rhetorical one. The price of gramophone disks fell sharply between 1903 and 1906, and the gramophone became a mass commodity in both the United States and Europe (de Meyer 1997: 48–51). No figures are known to underpin the claim that, proportionately, Dutch working class people owned more gramophones than middle and upper class individuals in 1913. Yet, there is evidence that in the 1930s, in terms of money spent, the gramophone, and, especially, the radio set, were relatively important to working class people compared to other forms of leisure. The most convincing evidence for this claim can be found in data collected by the Dutch Central Statistical Office (CBS) on the household accounts of almost 600 Dutch families in 1935 and 1936. The data show that working class families put "radio and gramophone" on top of the list of leisure items that they spent the highest proportion of their money on. In comparison, the middle and upper classes reported "holiday and trips" as the most important items of leisure expenditure (Centraal 1938: 20). Only a quarter of the workers did not have or use a radio, and having a telephone was more dependent on income than having a radio set. A trade magazine even claimed that low-income subscribers to radio services paid their radio fees earlier than their rents (de Wit and de la Bruhèze 2002: 355–356). In addition, the economic historian Henri Baudet has shown that in the 1930s,

Dutch working class families usually owned commodity radios, even though, unlike self-built ones, they were relatively expensive. From this Baudet deduced that the radio "had a very high priority" for working class families (Baudet 1981: 53–54). His study substantiates that the gramophone and the radio were of special *worth* and *significance* to the lower classes.

Although the progressives' resistance to the restrictions on gramophones and radios did not result in the rejection of the proposed ordinances, it did lead to changes in their formulation. In time, the grounds on which the loud playing of gramophones and radios could be legally punished became narrower. In The Hague (1927), Leiden (1928), Breda (1928), and Maastricht (1930), not the use itself, but the "bothersome" use of these devices was subject to fines. In The Hague and Leiden, the restrictions applied to the use of gramophones and radios both inside and outside private homes, but in Breda they referred only to noise that could be heard in the immediate environment of the devices being played. In Maastricht, The Hague, and Groningen, the authorities explicitly aimed at regulating noise violations produced by neighbors and on public streets, such as those made by shop-owners and radio manufacturers who tried to sell their products too loudly. In narrow streets, this kind of noise often caused traffic jams because people stopped to listen. The authorities in Maastricht thought their right to interfere in cases where noise could be heard only in private places (for instance at the back of a house) questionable. They decided only to exclude troublesome noise from radios and other mechanical devices that could be heard on public streets and only after violators had received a warning.[21] Such deliberations demonstrate the increasing difficulty city governments experienced in finding the legal grounds for intervention.

In the late 1920s and early 1930s, city councils in Utrecht (1929), Amsterdam (1931), and Groningen (1933) decided to regulate not only the noise nuisance coming from devices such as the gramophone and the radio, but also that from musical instruments. After the 1930s, this became a common policy in Dutch towns and cities.[22] Local authorities thus had begun to regulate traditional sources of sound anew while aiming to tackle the more recently introduced mechanical ones. This signaled that it was no longer considered acceptable, as progressive council members had argued all along, to make a legal distinction between the musical "noise" associated with high culture and that associated with low culture. Consequently, the specific character of the sound-producing *technology,* that is, that unlike classical musical instruments, gramophones and radios could be played incessantly, could no longer serve as the official grounds on which to base criteria for banning noise from neighbors.

That this was a remarkable shift becomes especially clear when we compare it with the campaigns against street music in Charles Babbage's times, as discussed in chapter 4. Babbage and his companions had displayed the same tendency to protect their taste of music at the expense of others' preferences as the proponents of restrictions on the use gramophones and radio had. To put it in Bourdieu's terms, they had enrolled their knowledge of good music—their cultural capital—to distinguish themselves from those who expressed a putative brutal taste, thereby enhancing their prestige and symbolic capital (Bourdieu 1989). In comparison, opponents to restrictions on street music had employed arguments very similar to those used by leftists in the gramophone and radio debates. They had claimed that "it was the working classes who enjoyed such amusements," and that interventions would thus act against their interests (Assael 2003: 191; Picker 2003). However, their arguments did not block interventions, whereas the ordinances on the nuisance of gramophones and radio gradually had taken the arguments on behalf of the working classes into account, and increasingly set limits to the bothersome playing of all kinds of music.

Complaints about the use of mechanical instruments, however, did not vanish. People still spoke of "radio din" and the "poisonous radio monster."[23] Yet very few violations were recorded under the new ordinances. Although data on the actual functioning of the ordinances against noise have been studied only for Leiden and Breda, the evidence suggests that, at most, only a handful of violations were recorded per year.[24] This could perhaps mean that the ordinances indeed prevented the abuse of radios and gramophones, or, that it was sufficient merely for policemen to mention the existence of the ordinances, as the city council of The Hague had hoped for.[25] Another possibility is that this kind of noise was not such a big problem after all. This is unlikely, however, given the persistence of complaints.

It is more probable that the ordinances were simply difficult to enforce. Some council members had feared this even before the ordinances had gone into effect, since it was far from clear how to decide whether any particular sound was "bothersome."[26] In 1937, the government of the Province of North Brabant sent a letter to the mayors of the other Dutch provinces to call attention to "the manifold occurrence of annoying sounds," including noise from musical instruments, loudspeakers, cars, and motorcycles. The letter urged local leadership to ensure that people adhered to the ordinances.[27] Three years later, in January 1940, the governor of North Brabant repeated the request, but now extended it to radio noise.[28] In response, Breda's deputy police chief explained to the city leadership that it was very difficult to decide whether noise had been audible only in the immediate environment of a radio or loudspeaker. He added that it might be a good idea to start reporting violations in "decibels," even though that "would require the purchase of expensive instruments."[29]

He may not have known that the use of measuring instruments in abating the noise of neighbors was also being discussed by the Amsterdam police, an issue I will address in the section "The Right to Make Noise," below. Before doing so, however, it is important to address another development that complicated the control of noise from neighbors by means of legal regulations.

The Pathetic Noise Complaint and the Problematic Pocket Noise Meter

Shortly after sound—whether live or generated by electrical devices—became legally understood in certain situations as noise nuisance, the notion of noise itself began to be disputed in new ways. Scientific studies from the mid- and late 1930s on the physiological and psychological responses to noise contributed significantly to such debates, turning noise nuisance into an even more complicated concept than before.

Since the late 1920s, scientists in Great Britain and the United States had been quite concerned about the effects of noise on man. As discussed in chapter 4, their research showed that experimental subjects experienced an increase in their rate of breathing and systolic blood pressure, and a reduction in the efficiency of their bodily and mental processes in response to noise. Such difficulties were especially relevant to neurotic city dwellers, the Noise Commission of London claimed, since "sleep is indispensable to the neurotic, who does the work of the world," and since the most disturbing noises for people who wanted to sleep were the sounds of the city. These sounds included those of "unusual and sudden horns, exhausts, drills, vibrations, whistles, and milk can deliveries" (Brown et al. 1930: 17 and 106–107).

From the mid-1930s onward, however, experts on psychoacoustics increasingly stressed that the "reaction to noise is largely temperamental and varies greatly from person to person, and that for the same person it varies with the conditions under which the noise is heard" (Davis 1937: 10). Everyday life had always shown that "some persons complain of noises which others find innocuous" (Davis 1934: 260). Yet little was known about the conditions under which "some persons" felt disturbed by noise. The experimental psychologist Bartlett explicitly addressed this concern In *The Problem of Noise*. He wanted to qualify "the menace of noise" (Bartlett 1934: 1). He believed that serious damage to hearing occurred only in special occupations, such as those of boilermakers. The various noises to which the average citizen was exposed were generally not loud enough to impair their hearing. Other scientists discussed the "transient" physiological effects of noise that "normally pass away rapidly and completely" (Bartlett 1934: 18).[30]

Bartlett also criticized the putative effect of sound on the performance of work. On the basis of his own and others' experiments, he argued that what mattered was not the noise per se, but the noise in relation to the task. "On the whole the more the work puts a demand on the higher mental processes, the more disturbing is the noise likely to be." Whether or not a sound became treated as a nuisance depended on the background against which the sound was experienced, its loudness, it ambiguity of direction, its (dis-)continuity, and its (non-)familiarity. Moreover, the "disturbing effects of noise are at their maximum for people who have to do mental work, but are for some reason bored, tired, forced into a job which is a bit too difficult for them or not quite difficult enough, or in which they are only moderately interested." Bartlett even suggested that the "complaint against noise is a sign, sometimes, of a deeper social distress" (Bartlett 1934: 53). As another scientist recapitulated, "In general, when a break-down in health occurs under exposure to noise, there are other influences at work. The noise seems to act as a catalytic agent or accessory factor, thereby inducing or accentuating a nervous state" (McLachlan 1935: 134).

This subjectification or even *pathologization* of complaints about noise by *scientists* was new. Before, it had been primarily intellectuals who had publicly underlined the need for silence. Their antinoise essays and pamphlets were rife with notions of a civilized battle against barbarous noise, of silence being "distinguished" and noise being brutal. Their campaigns had pitted intellectuals against workers, or sensitive individuals against indifferent ones. Some of their opponents had ridiculed their complaints as hysterical. Yet now experts from the rising field of psychoacoustics challenged the views of the intellectual elite. Noise was not simply produced by "the other"; it was also a product of one's state of mind. This new notion of the subjectivity of sound perception made it increasingly difficult to decide which sounds could or could not be treated as a nuisance. It had previously been acknowledged that some individuals were more sensitive to noise than others, but such sensitivity was no longer considered a mark of social superiority—rather, it was viewed as something problematic: the result of a troubled personality or a strained mental state. In addition to the difficulties involved in enforcing the ordinances governing the use of musical instruments, these experts' new stress on the subjectivity of sound perception further complicated legal interventions into issues of neighborly noise. Given such setbacks, it is quite understandable that local police officers longed for solutions in terms of decibels.

Acousticians had been quite optimistic about the establishment of quantitative criteria for the definition and control of noise. As scientist E.E. Free claimed in 1930, the "acoustic expert may be called upon to measure as noise anything from the neighbor's piano playing to the crash of thunder" (Free 1930: 19). And indeed, the decibel helped to concretize the results of architectural acoustics in excluding noise from buildings. Through instruction pamphlets, lectures, exhibitions, papers, and books on "the silent house," acoustical engineers explicated what "resonance" could do to buildings, what the differences were between air-borne and structure-borne sound, and what such differences meant to the issue of noise abatement in buildings. They talked about establishing criteria for sound reduction, sound transmission, sound absorption, and sound insulation and about the materials fit for excluding air-borne sound, for, as they believed, the velocity of sound propagation was dependent on the specific gravity of materials. The 1935 London Noise Abatement Exhibition displayed not only products reducing the noise of cars and other forms of transportation, but also floating floors, sound-deadening partitions, and products such as a "silent syphonic w.c. suite," silent vacuum cleaners, and nonslam automatic doorstops (Noise 1935). (Figure 6.1.) What's more, acoustical engineers and architects showed how cavity walls, double windows, and specific doors could diminish the noisiness of houses expressed in dB.[31] (Figure 6.2.)

Yet, police officers' dreams about instruments enabling the registration of violations of ordinances governing the use of musical instruments had probably been more directly inspired by what had already been achieved by measuring traffic noise in the Netherlands and abroad. As they would soon find out, however, chasing noise offenders with the aid of noise meters on a day-to-day basis was far from easy. In 1933, the Dutch conservative national daily newspaper, *De Telegraaf,* bolstered the noise abatement effort by inviting Harry P. Samuel, an engineer from Western Electric Holland, to ascertain the nuisance of traffic noise with help of a "silent witness," the noise meter (Anti-Lawaai-Campagne 1933). One article featured Samuel taking notes of "the wild movements of the needle" in the presence of leading police officials (Eerste aanval 1933). Other articles discussed the "hunt for decibels" and the levels of street noise coming from streetcars, trains, motorcycles, lorries, carts, automobiles, and automobile horns (Geluidsopmetingen 1933; Nieuw 1937).

In 1937, the Amsterdam chief of police C. Bakker and acoustical engineer professor C. Zwikker joined forces in the creation of a simple, low-cost pocket-size noise meter to be used by police on the streets. Bakker had argued for introducing maximum noise limits, and after their legal formulation in the Motor and Cycle Regulation of 1937, he needed a noise meter to enforce them. All of

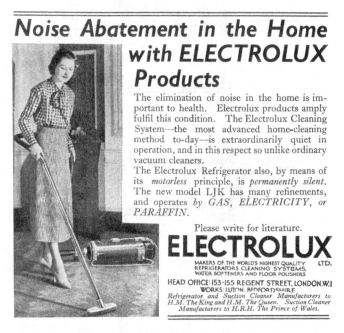

Figure 6.1 Advertisement for quiet ELECTROLUX products, 1935.
From *Noise Abatement Exhibition* (London: The Anti-Noise League, 1935),
p. 74.

the available meters, however, were either too large, too expensive, or too sensitive. Bakker felt that while they might be suitable for use in a laboratory, they were altogether impractical for use by police officers engaged in tracking down noise offenders on the city's streets. He therefore asked Zwikker to develop a more practical device that would suit his particular needs (De Geluidsmeter 1937).[32] Zwikker—or actually, one of his students, F.W. van Gelder—did so, which resulted in the pocket noise meter named the Silenta. The Silenta was able to measure noise levels only above 80 dB, which was acceptable because the 1937 regulation had set a maximum of 95 dB for noise made by car horns, and a maximum of 85 dB for noise produced by automobiles and motorcycle engines. In 1939, the meter was accepted as a legally valid measurement device

So

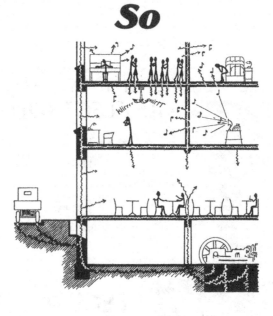

oder So?

Figure 6.2 *So oder So* (So or So)? Illustrations of appartment buildings with and without noise transmission. From a publication on "the silent house," issued by the German Association of Engineers, 1934. From *Das lärmfreie Wohnhaus,* (Berlin: Verein Deutscher Ingenieure, 1934), pp. 86–87. Courtesy Verein Deutscher Ingenieure.

by the Dutch Supreme Court (Zwikker 1972: 92–93; Zwikker 1939/1940; de Bruijn 1984: 50).

The process of certification required for this acceptance had proceeded all but smoothly (Nieuw 1937; Uitspraak 1937; Zachte 1938). Yet, even larger problems had to be overcome in the everyday use of the pocket noise meter. The Amsterdam Police began using the Silenta in 1937, as the "mechanical ear" of a newly established unit of the Amsterdam Police, the Silence Brigade. Under supervision of a police inspector, four policemen equipped with a motorcycle with sidecar were deployed to abate excessive city noise. Their main task was to reduce the noise from unnecessary, polytonal, or voluminous sound signals and badly constructed exhausts, as well as to educate drivers not to sound their horn when they approached an intersection.[33] The Silenta's readings enabled the officers to provide drivers with evidence of their offense. In fact, after some time, the Silence Brigade discovered that automobiles and motorcycles only rarely exceeded the maximum sound level. It therefore proposed that the maximums be lowered to 85 dB and 75 dB. Their proposal was partially implemented in 1939, when the Motor and Cycle Regulation fixed the new maximum levels at 90 and 80 dB.[34]

According to Zwikker, however, the Silenta failed to become a commercial success. Apparently, local authorities did not buy the pocket noise meter as often as Zwikker had expected. He attributed its failure to the advent of the Second World War and the ensuing shortage of repair parts (Zwikker 1972: 93). Whatever role the war might have played, the truth was that the Silence Brigade experienced drawbacks in the day-to-day use of the device. It was common for members on patrol to stop a motorist as soon they heard a suspicious noise. As journalists testified in their articles, the members of the Brigade had well-trained ears for potential offenders; on their motorbikes they skillfully—albeit not silently—chased rowdy suspects amid the chaos of traffic. Problems sometimes arose, however, after stopping motorists. According to the regulation, a vehicle's noise had to be measured in "open space," at a distance of seven meters. Given the many narrow streets of Amsterdam, this stipulation did not prove very practical. Both police reports and the press described how the Brigade tended to deploy the Silenta on whichever street a potential offender happened to be.

If the readings surpassed the maximum level, the officer then had to ask the motorist to follow them to an open space where the vehicle's noise would be measured again. In most open spaces, however, the almost ever-present wind could dramatically affect the Silenta's readings. The discrepancy between the first and second readings the winds created, depending on their force and direction, could be as much as 17 dB, enough to disqualify the readings' validity (Razzia 1937; Met 1937; Actie 1937; Taxi's 1937; Geluidsoverdaad 1937; De stiltebrigade 1937).

Other clients who used the Silenta claimed to have experienced similar difficulties. The Silenta gave readings that tended to be considerably different from those produced by a similar device made by General Radio, and its sensitivity to wind when used in the open air made it inaccurate. Zwikker cautioned users to be aware of these problems, but claimed that they did not negatively affect the police's use of the Silenta. Moreover, he explained that his noise meter was calibrated for a mix of tones unlike the General Radio device, which could read only pure tones.[35]

Zwikker may not have been wholly forthcoming about the Silenta, however. As will be shown in the following section, his pocket-size meter also became a controversial measuring tool in the context of neighborly noise. Clearly, then, the effort to objectify noise by means of noise meters—in a situation in which experts from psychoacoustics increasingly stressed the subjectivity of sound perception—was accompanied by several challenges. First, measuring street noise was one thing; creating effective or reliable pocket-size meters for police use was quite another. Second, it proved difficult to deploy the meters in quarrels about noise between neighbors.

THE RIGHT TO MAKE NOISE

The Silenta was designed for measuring street noise. Such noise also constituted the focus of the Amsterdam silence campaign in the late 1930s. Yet, according to Police Chief Bakker, the noise produced by radios did not fit into the category of street noise. He claimed that the trouble with radio noise was that its production could be ascertained only *inside* private homes. Under normal circumstances, the police were solely allowed to enter a private home after a neighbor made a complaint. This spatial restriction governing intervention conflicted with Bakker's preference for an active, sustained police campaign against noise.[36] He felt that at a later stage of the campaign it might prove necessary to tighten some regulations. However, before the issue came up for discussion among those directly involved in the campaign, voices in the press had already articulated views about the import of Bakker's mission that were emphatically unambiguous.

The potential deployment of the Silenta in cases of domestic noise nuisance was ridiculed in *De Telegraaf,* the same newspaper that had initiated the noise abatement campaign. "Amsterdam has its 'Silence Brigade,' which will take action against all those within the city's confines who make needless noise," proclaimed one author, who chose to remain anonymous, in a column published July 7, 1937. It is "a great measure," he added ironically. Imagine, he told readers, that you hear the twin babies of your neighbors screaming: "You snatch the phone and dial our silence dictator, Police Chief Bakker, and explain the case to him. . . . Hardly three minutes later, you can hear the silence brigade's motorcycle rush into your street. . . . With them, they have professor Zwikker's sound meter and indeed, true enough—this *is* too much needless noise. Yet whereas the noise little John produces exceeds the limit by 0.12, little Peter, his twin brother, exceeds the maximum limit by no less than 8.65 dB. What is the brigade to do in such a case? Arrest little Peter and take him with them, regardless of the protests of his happy parents?" After telling stories about other fictitious cases, the author concluded that the brigade might well run out of "zwikkers"—a nickname for the Silenta (Spotternij 1937).

If this was not enough, a year later *De Telegraaf* published another fictional dialogue between two neighbors: "By the time I wanted to read my newspaper serial last night, you were playing dance music, madam, as many as eighty-three *zwikkers* too loud. And somewhat later I could hear that in your kitchen you were speaking your mind to Trude, who first dropped the tureen—sixteen decibels too loud—and subsequently talked back to you, thereby largely exceeding the legal *zwikker* limit.... Family life will become increasingly peaceful, since as soon as a particular family member raises his voice, he will not be responded to somewhat more loudly but one simply fetches the *zwikker* and disarmed this family member will be" (Spotternij 1938). By exaggerating the precision and potential advantages of the noise meter in domestic and neighborly conflicts, the author of the mocking article was suggesting that some noise should merely be accepted, and that any effort to objectify noise without taking into account the context and meaning of sound was a completely ridiculous effort.

In these situations, the author submitted, the limitations of the sensible use of the noise meter were utterly clear from the start. Not a peaceful life, but a world of conflict would be the result of deploying a technology like Zwikker's noise meter in conflicts between neighbors. A truly calm neighborhood could be secured only if people accepted each other's right to make noise to a certain extent, and refrained from calling the police for every minor incident. As early as 1933, *De Telegraaf* had approvingly quoted a Paris police prefect who said that it was necessary to compromise between people's need for silence on the one hand, and their wish to listen to newsreels on the radio and their love of music on the other. To his mind, "city life" could not do without a certain degree of "courtesy" (Het groote-stadsleven 1933). The mockery of the potential usefulness of Zwikker's *Silenta* in quarrels between neighbors was perfectly in line with this reasoning.

The historian Raymond Wesley Smilor claims that a comparable attitude also prevailed in America. Because everyone lived within a community, most individuals would be hesitant about grumbling over the noise made by others, because they too would not want any one to gripe about the noise they made. Most people would vacillate before "invoking such powers of the law as are clearly his," since they would not want their neighbors to consider them "a kill-joy or opposed to 'progress.'" In every community, "the rule of 'live and let live'" held "considerable sway" (In the Driftway 1929, quoted in Smilor 1971: 25). The examples used by the Dutch *Telegraaf* even suggested fears of aural surveillance, which may, again, have hampered the acceptance of employing noise meters in neighborly quarrels about noise.

In addition to the argument that the masses had a right to use their gramophones and radios and that sound perception was highly subjective, the argument that residents had a right to make noise at home significantly influenced post–World War II discourse on the noise of neighbors. Taken together, these three arguments made it increasingly unlikely that the noisy behavior of neighbors would become regulated as part of a national Noise Abatement Act, let alone that it would be controlled by quantitative criteria.

Experts in acoustics, psychoacoustics and social medicine who prepared reports on this issue, whether for their own institutes or for national authorities, kept stressing the variety of an individual's sensibility with respect to noise. A Dutch survey on the annoyance of the sound made by radio, talking, walking, trampling, vacuum cleaners, and children playing, published in 1958, claimed that "those who did intellectual work usually experienced more sound nuisance than those who worked with their hands." One's sensibility for sound nuisance tended to go up as one's education, income, social standing, children's age, and the level of intellectual activity (as well as that of one's family members) increased. In contrast, "sound nuisance showed a tendency to decrease when the size of the family increased" (Bitter and Horch 1958: 85–86). The experts stated again and again that responses to noise were strongly influenced by one's personality and situation as well as by one's mental and physical condition (Bitter 1960: 206–207; Geluidshinder 1961: 15, Geluidhinder 1965: 2; Bitter 1967: 325). One of the most prominent of these experts, the psychologist C. Bitter, cited Bartlett's book and other Anglo-American research to underpin his claims (Bitter 1960: 213).

Apart from acknowledging individual differences, social scientists anticipated a growing need for leisure—which some people enjoyed *with* loud sounds and others *without* them. Given this clash in tastes, they had reason to project a sharp rise in problems associated with noise nuisance. If the sound insulation used in constructing housing was not improved, they argued, it could lead to a pattern of life that would frustrate any form of spontaneity, since most people were afraid to disturb their neighbors, which, in turn, would hamper their own mental development or the family's well-being (Bitter 1960: 209). In the

181

domestic context, then, individual variety in sensitivity to noise complicated the noise problem even further. It was taken for granted that people somehow needed to express themselves within their homes. "Although I do not want to maintain that one should not be a little more considerate of other people," an expert in social medicine wrote, "one should not forget that making sound, at times even much sound, is a very normal manifestation of life" (van den Eijk 1960: 277). In the years following this claim, he said that listening to a good record might be as pertinent to one's personal development as reading a good book (van den Eijk 1969: 18), and even asserted that drawing attention to oneself by making noise was an inevitable phase in a person's development (Geluid-shinder 1961: 3).

To be sure, authorities were not entirely powerless. From the late 1930s onward, acousticians advocated not only good city planning, but also the reduction in the emission of sound from domestic appliances as ways of solving the noise problem at home. In addition, the first provisional norm for soundproofing in housing was articulated in the early 1950s.[37] In 1966, after years of research in test buildings, the Dutch government declared the NEN 1070 (1962) to be in force, which replaced a provisional norm (Voorschriften 1966).[38] Intriguingly, this norm for new housing incorporated the idea that soundproofing should be of such a quality that residents would "hardly" be disturbed by the radio of their neighbors—at a period in time in which television and stereo sets increasingly appeared in living rooms (Geluidshinder 1961: 6; Geluidsisolatie 1961: 212; de Bruijn 1984: 43–44).

Toward the end of the 1960s, social scientists claimed that people had become more susceptible to noise because of their hurried way of life. In view of this new lifestyle, they underscored the significance of silence and tranquility. Such an environment would help people regain their strengths, which in turn would be of benefit to their family life and professional output. However, urban life in particular was full of ambiguities. People looked for amusement as well as tranquility, and the former always seemed to be accompanied by some form of sound. Some experts even observed that city dwellers' habituation to noise could be such that many were no longer capable of sleep in a quiet village setting anymore. The fact that cities had grown into more densely populated ensembles and more and more urban residents lived in apartment buildings, while at the same time culture at large generated more differentiated patterns of activities, also contributed to the emergence of new dilemmas associated with neighborly noise in urban settings (Bitter 1967: 327–328; Schmidt 1969: 4).

The acoustical engineer C.W. Kosten suggested a way out of these dilemmas. He took issue with those who believed it to be impossible to do something against noise nuisance because of the subjectivity of sound perception. Kosten claimed that one should always look for a combination of objective noise measurements and intersubjective judgment. Especially if one only bothered about addressing noise excesses, or those cases that the vast majority of the "decent-minded" public thought were unacceptable, such an approach would work well (Kosten 1969: 34). The criminologist Wouter Buikhuisen disagreed. In an essay about Dutch beatniks and their preference for revved-up mopeds, he contended that noise abatement would fail to be successful if one did not bear in mind that sounds had a specific psychological or cultural meaning to the people who produced them. He proposed two alternative solutions: diminish the level of nuisance by helping those bothered by noise to grasp the situation intellectually, or provide those who cause the nuisance with an acceptable substitute. The display of bravura had been part and parcel of youth culture for ages, and therefore any effort to repress it would lead nowhere (Buikhuisen 1969: 8).

It is unlikely that most experts at that time subscribed to Buikhuisen's empathetic approach to moped-loving beatniks. Yet the notion that making a certain amount of noise was a basic right of residents continued to prevail. In light of this view, the Dutch Board of Public Health, which advised the government on the issue of noise and which retained Bitter as an adviser, defined acoustic privacy in two ways. Acoustic privacy was the right to an undisturbed tranquility inside one's home *as well as* the right within the confines of one's home to express oneself freely, uncontrolled by others.[39] Accordingly, it proposed to address problems involving residential sound nuisance through four measures: soundproofing, reducing the sound emission from consumer appliances, careful city planning, and public information campaigns.

Hence new national interventions in and stricter criteria for the regulation of the noisy *behavior* of neighbors through penal law was an unlikely course

183

of action. In 1975 the Dutch government sent a preliminary version of its first national Noise Abatement Act to Parliament. In the wake of a growing awareness of environmental problems, noise was put on the national agenda with new élan. The Noise Abatement Act had prescribed regulations against several kinds of noise, including industrial noise, motorized traffic, trains, and discos. Many of these regulations were based on maximum levels of sound emission. Yet the act had far less to say about noise nuisance in the domestic, or residential, sphere. Although the act's explanatory notes made clear that noise from neighbors was a source of annoyance for one out of four citizens, and although the act set standards for the use of sound absorbing walls in housing, the government was unwilling to issue any national rules against the noisy use of equipment in private homes.[40]

Initially, members of parliament and spokespersons for noise abatement and environmentalist organizations expressed their surprise about this reluctance. The "industrialization" of homes, they argued, had brought complete audio systems into living rooms, effectively turning them into concert halls. Did not many people suffer from neighbors who refused to lower the volume of their stereo set or use it only during certain times of the day?[41] The answers that Minister Irene Vorrink gave were quite ambiguous. On the one hand, she stressed that the noisy behavior of neighbors should not be part of national law, but of local ordinances. On the other hand, she claimed that such ordinances were not effective, because they were difficult to monitor and uphold. It was far from easy for the state to interfere in people's private homes. Moreover, people's standards differed; what some considered normal, others saw as nuisance. Therefore, in practice it was more or less impossible to gain control over neighborly noise in legislative terms.[42] What the government proposed instead was subsidizing a huge information campaign, entitled "Let's Be Gentle with Each Other," which would be organized by the Dutch Foundation for Noise Abatement (*Nederlandse Stichting Geluidhinder* [NSG]), established in 1970.[43] This proposal eventually passed the Dutch parliament.

The focus of the information campaign "Let's Be Gentle with Each Other" was clear. (Figure 6.3.) It started from the assumption that everyone decides for himself or herself "whether and when noise is annoying" and that "each

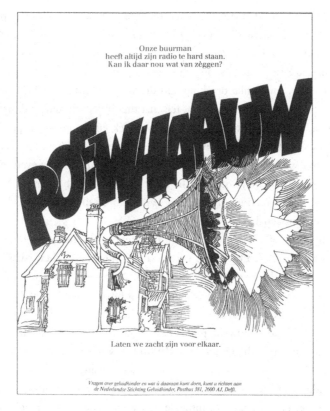

Figure 6.3 Advertisement for the Dutch campaign Let's Be Gentle
with Each Other, late 1970s, which says "Our neighbor always plays
his radio too loud. Can I say my piece about it? Let's be gentle
with each other." From *The Soft Sell. Making noise a public campaign,*
(Amsterdam: Instituut voor Sociale Kommunikatie, 1979), p. 7.
Courtesy Eric Ezendam.

individual sets other standards." Therefore, "the only way to solve this problem is to work it out together" as a community (The Soft 1979: 16). An early publication of the Dutch Foundation for Noise Abatement advanced the view that producing excessive sound should become a phenomenon people would feel ashamed about, much like drinking and driving (Geluidhinder 1971: 24). The designers of the campaign, however, felt that the subject should be dealt with in a "prudent, understated way" and that any shock effect, insult, or paternalism was to be avoided. Their final objective was a modest one. They wanted to "create a positive willingness to think about the subject. And maybe even to realize that once in a while they could be the source of another's bother.... Again: no reproaches, no corrections. For people will tend to shy away from those and become unapproachable." (The Soft 1979: 5–6).

One series of ads in the campaign used texts written from the perspective of individuals bothered by noise, which said things like, "Their party is noisy now. But our dog keeps them awake, at times. With some good intentions from both sides, such problems should be properly soluble" (The Soft 1979: 20). The pamphlets promised that talking with one's neighbors would help, and this was highly preferable to calling the police (The Soft 1979: 29, 35). In the years following the campaign, the Dutch Foundation for Noise Abatement promulgated suggestions for material reductions of noise, such as using headphones or closing one's windows, but also advised citizens on how to talk with neighbors who kept making noise. They should, for instance, not wait to address the neighbors until the noise made their blood boil, but should choose "the right moment" for "a calm talk" (Geluidhinder [1992]: 4).

Initially, the Dutch Foundation for Noise Abatement claimed that the campaigns were quite successful in raising people's awareness of the problem and increasing their willingness to do something about it. With respect to actual changes in people's behavior, however, the Foundation later had to acknowledge that the campaigns were far less successful. In 1990, over a decade after the beginning of the campaign, only 45 percent of the Dutch populace had enough courage to talk to their neighbors when they felt bothered by their noisy behavior. Even worse, the number of cases in which such discussions actually produced a positive outcome declined significantly from about 65 percent in 1990 to 33 percent in 1997 (Effecten 1997: 2–3). As early as 1967, Hans Wiethaup, member of the German Supreme Court, discussed similar problems in Germany. Although German citizens could officially refer to the civil code, just as in the Netherlands, they hardly did so out of fear of addressing their neighbors in such a direct manner (Wiethaup 1967: 86).

CONCLUSIONS

In this chapter, I have focused on the debate about noise, nuisance, and neighbors as it materialized in the Netherlands during the course of the twentieth century. I argued that the decision *not* to control the behavior of neighbors by means of science and technology-based legal criteria resulted from three significant changes. The rise of a particular political climate, of new expert knowledge on the subjectivity of sound perception, and of the scientifically and politically acclaimed right to make noise in one's home all contributed to the evasiveness of the public problem of neighborly noise.

The noise nuisance that followed in the wake of the increasing availability of gramophones and radio sets in the first decades of the twentieth century initially came up for debate in urban neighborhoods, where a substantial number of residents lived in close proximity, often in modern-style concrete apartment buildings that were all but soundproof. The residents of these neighborhoods embraced various lifestyles—some working shifts in factories, others actively engaging in the city's nightlife—making city residents more vulnerable to noise nuisance than others. Furthermore, experts from acoustics and psychoacoustics associated the problems of noise with urban life in particular because city-dwellers were seen as the main contributors to the country's economy. Since they were already exposed to a multitude of sensory experiences throughout the day, they badly needed their sleep.

When confronted with the noise problem generated by the mounting presence of gramophones and radio sets, local Dutch politicians focused on the infinitude of the sound of these machines, and wanted to issue strict regulations for their use. Soon, they were forced to take nonmechanical sound, such as that of singers and musicians practicing scales, into consideration as well. Leftist council members emphasized the class issue involved, since they approached gramophones and radios as the musical instruments of the masses, and not without reason. It was considered unfair to single out the nuisance of loud gramophones and radios and control their use by means of local ordinances, without also regulating nuisance caused by traditional sources of sound, such as musical instruments. What the working class communities did, as the progressives saw it, was sharing sound rather than intruding in other people's privacy. The effort to establish a formal distinction between technology-induced noise nuisance, as that made by gramophones and radios, and noise nuisance caused by musicians, ultimately failed. The resulting local ordinances aimed at banning noise nuisance in city neighborhoods were difficult to uphold, though, in part because they implied the active interference of local authorities in people's private domain.

Moreover, psychoacoustic experts established that one's reaction to noise was largely temperamental. This change in how persons reacted was very significant. It was not that intellectuals had never before been aware of individual differences in the sensibility to sound. On the contrary, this had been a theme for generations. Yet being sensitive to noise acquired a completely new status. Once a sign of one's level of education and refinement, sensitivity to noise was now seen as a sign of not being able to cope with one's mental work, of deep and pathological distress. And once a legitimate cause for the elite's control of sound, it became an argument for underlining the difficulty of intervening in noise issues.

This did not exactly help in finding a legal basis for abating neighborly noise. Police officers, however, hoped to find solace in the use of the noise meter. Inspired by the successful measurement of traffic noise in New York and elsewhere, and seeking to execute traffic noise abatement laws of the late 1930s, the Amsterdam Police Department showed an active interest in the development and application of a pocket-size noise meter. Tentative suggestions to employ noise meters in neighborly conflicts were formulated as well. The Silenta, the custom made measuring device produced as a result, had clear limitations, however. The wind as well as the reverberations of buildings along city streets could substantially reduce the reliability of the pocket noise meter's readings. Furthermore, the noise meter's potential for solving conflicts about noise between neighbors became the object of ridicule in the local press. Journalists argued that such conflicts, which formed an inextricable part of residential life, called for courtesy rather than noise meters. More generally, experts increasingly began to stress people's right to make noise in the domestic sphere. Practicing music or listening to it at home was widely viewed as an important form of leisure, just as the freedom to express oneself counted as a crucial aspect of privacy. Consequently, there were fewer reasons to believe that a political effort to regulate the domestic production of noise formally on the basis of quantitative criteria could be successful.

It is important to note that these developments were not confined to the Netherlands. In the early twentieth century, the sound of gramophones and radio sets became the subject of heated discussion all over the West. The growing conviction that sound perception had a large subjective element to it was clearly an international academic trend. The broadening of local ordinances to cover all types of domestic noise, including that of musical instruments, also seems to have been a rather general development. In Paris in the early 1930s, for example, an ordinance was established banning nuisance produced by gramophones and radios as well as musical instruments. And Likewise, during the same period, the New York Noise Abatement Commission sought to broaden an ordinance against the "excessive and unusual noise" of mechanical or electrical instruments into one that included musical instruments as well.[44] In Germany, authorities had great difficulty finding the legal grounds for intervention. As early as 1915, however, the courts decided to recognize local bans on making music as well as playing gramophones in front of open windows in the evening and at night (Saul 1996b: 169–170).[45]

In England, the Home Office prepared a model regulation that specifically prohibited nuisance from "loud-speakers" as late as 1938 (The Neighbour's 1938). A year later, the Noise Abatement League proposed to suspend the radio licenses of those who disturbed their neighbors. Yet a Daily Mirror cartoon immediately questioned whether this would work (Editorial Notes 1939; Will 1939). Moreover, under the British Noise Abatement Act of 1960, noise nuisance "from any source" could be an offense. And even more illuminating is what the famous 1963 Wilson Report on Noise stipulated: "Ultimately, the best remedy" for "the disturbance from neighbours' radio and television sets, record players, and tape recorders" was "sympathy and consideration between the people concerned" (Wilson 1963: 117; Government 1962: 18; Noise 1960).

These findings point to the significance of the timing of noise abatement. In the nineteenth-century fight against stench, authorities still felt fully entitled to intervene in private households and had a police force strong enough to help them. In twentieth-century efforts to abate neighborly noise, they did not feel fully entitled and, at times, equipped to do so. In the nineteenth century, the sensitivity the higher classes displayed about foul smell helped to legitimize regulation of household noise. From the late 1930s onward, the sensitivity of intellectuals toward noise itself began to smell bad, and could thus not be used as legitimation to those wanting to tackle the noise of neighbors. In fact, the noise and neighbors debate—and notably the introduction of the right to make noise—points to a turn of the tide in "acceptable" behavior and the relations between classes that would be far reaching. This will receive a more elaborate discussion in the conclusion of this book. For now, it is important to note that even the technologies dominating this chapter—the gramophone, the radio, and the noise meter—acquired a new status over time. The mechanical musical instruments transformed from machines producing endless sound that thwarted people's privacy into instruments suitable for sharing sound, expressing one's personal development, and handling the tensions of modern life. And the noise meter became less reliable and less widely usable than initially thought.

Apart from the still existing penal and civil code sections, the noisy behavior of neighbors remained "beyond" law, and the trust in numbers that prevails in so many domains of modern society did, apart from the norms for sound insulation in housing, not lead to a regulation in terms of decibels. In contrast, authorities hired experts from psychology and related fields to educate citizens to solve the problem of neighborly noise among themselves. For the historian of psychology Nikolas Rose, this may not come as a surprise. According to him, psychologists, "the engineers of the human soul," have increasingly been given a role in dealing with societal problems—through the kind of public education as in "Let's be Gentle," for example—because today's liberal democratic forms of government ask for rationality, privacy, and autonomy. The exercise of power needs to be justified on a rational basis, and thus on expertise. At the same time, liberal democracies guarantee its citizens autonomy and private space, "outside the formal scope of the authority of public powers." Yet what happens within these cocoons of privacy is considered vitally significant for "national wealth, health, and tranquility." Psychological expertise opens such private space without breaching it, thus reconciling the need for intervention and privacy. It even constructs "a kind of regulated autonomy for social actors," a promise of "competent autonomous selfhood" for all (Rose 1992: 367). In terms of the noise of the neighbors' problem: every citizen was entitled to a sophisticated education in how to accept some noise, or how to delicately convince his neighbor of the seriousness of his problem. This, in fact, not only partially dequantified, but also individualized and depoliticized the process of dealing with the noise of neighbors, and at a moment when the right to create noise made noise problems between neighbors increasingly likely.

I will discuss the implications of this situation in the next chapter. The outcome of the noise-and-neighbors debate did not mean the end of a trust in quantitative criteria in other noise issues, however. The abatement of aircraft noise, for instance, would be full of numbers. Let me thus first trace the booming business of aircraft noise.

A Booming Business:
The Search for a Practical Aircraft Noise Index

A European Harmonization

In 2000, at the start of the new millennium, the European Commission presented a new approach to the management of environmental noise. It aimed to "harmonise noise indicators and assessment methods for environmental noise" and to use these standards for making "noise maps." Such maps, once made public, would become the basis for action plans at the local and EU level.[1] If harmonizing noise indication and its effects was seen as a first basic step toward developing a new European noise abatement policy, most of the existing noise indexes were considered to be insufficiently adequate to function as a common standard.

A noise indicator, according to the European Commission's definition, was "a physical scale for the description of environmental noise" that had "a demonstrable relationship with a harmful effect."[2] The new standards, *Lden* and *Lnight,* were designed to indicate not just one type of noise, such as aircraft noise, but environmental noise in general, including road traffic. The proposed indicators departed from many of the EU member states' aircraft noise indexes, such as the British *Noise and Number Index (NNI),* the German *Störindex (Q),* and the Dutch *Kostenunit (Ke).* The British introduced their *Noise and Number*

Index in 1963, while the *Störindex* of the Germans went back to the mid-1960s, and the Dutch adopted their *Kostenunit* in 1967. Many of these regulative standards had already been under attack for some time in their respective national contexts. Now, the European Commission intended to eliminate most of the national indexes in one blow.

Yet why did the various European countries have distinct aircraft noise indexes in the first place? When by the late 1950s the issue of aircraft noise became a more prominent public concern, acousticians explicitly defined this type of noise as an international problem that called for a standardized, supranational approach. As such, the aircraft noise problem seemed to constitute a perfect example of what public problems' theorists have dubbed the "social construction of similarity" (Best 2001: 11). These academics claim that public problem definitions and solutions travel most smoothly from country to country in cases in which these states experience similar socioeconomic and technological trends, share languages and cultural beliefs, and interact through international professional networks in which experts theorize the problems at hand (Best 2001). From this angle, it is hard to understand why Europe has had such an array of national aircraft noise indexes since the 1960s. The rapid expansion of aviation was a widespread Western affair, while the field of acoustics had a genuinely international focus ever since the late nineteenth century (Ku 2006: 407). Knowledge of aircraft noise indexes circulated freely from country to country. Still, the British, the Germans, and the Dutch chose to formulate distinct standards.

This chapter aims to understand the proliferation of the various national aircraft noise indexes and their consequences for both national noise abatement policies and recent attempts in Europe to harmonize them. It shows how acousticians became simultaneously involved in national policymaking and in seeking an international agreement, and it discusses how long-standing research and legislative traditions affected their local decisions, which, in turn, retarded a supranational approach. The complex relations between the international organizations dealing with aircraft also had a decelerating effect on international standardization. Yet this chapter underlines, again, how the framing of new noise problems drew on older ways of dramatizing, defining, and tackling noise problems.

This is only part of the story, however. The experts who pursued a practical aircraft noise index were also struggling with the tensions between the ideal of mechanical objectivity and the logic of what I will call *pragmatic objectivity*. While mechanical objectivity stands for trust in the disembodied and transparent quantification of phenomena, pragmatic objectivity refers to the logic of intervening on the basis of numbers. Whereas mechanical objectivity aims at showing the manifestation of a particular phenomenon in its full scale variability, pragmatic objectivity is geared to distinguishing between sharply demarcated levels of the phenomenon's manifestation that can be linked to discrete degrees of intervention. The tensions between these two forms of objectivity contributed, I claim, to the proliferation of national aircraft noise standards. The ultimate result was a series of standards that were officially transparent, created strictly defined zones with distinct levels of allowed noise, and had significant consequences for citizens, but were largely beyond their grasp and immediate control.

This chapter closely analyzes the development of aircraft noise indexes in the United Kingdom and the Netherlands. Where relevant for my argument, I will also discuss the origins of aircraft noise indexes and norms in other Western countries, such as the United States and Germany, albeit in less detail. I foreground the situation in the United Kingdom because in the late 1950s, its main airport, London Heathrow, was by far Europe's biggest airport in terms of the number of starts, departures, passengers, and tons of freight handled (Meister 1961: 46; Bouwens and Dierikx 1996: 130, 141). It was also surrounded by densely populated neighborhoods that were increasingly exposed to the highly disturbing sound of new machines: the screaming of the turbojet airplanes that succeeded the last generation of propeller-driven, piston-engined aircraft. American jets arrived at Heathrow from 1958 onward. This turned Heathrow, as a German report claimed, into "a kind of test case for the exposition of the aircraft noise problem" (Bürck et al. 1965: 97).

In 1957, Amsterdam's Schiphol Airport was only sixth on the list of European airports with the most starts and departures (Meister 1961: 46). What makes the Netherlands an informative site of study, however, is that the Dutch noise experts deliberately departed from the British standard, even though they fervently advocated an international approach to the aircraft noise problem and had closely followed deliberations over the issue in the United Kingdom. Thus a consideration of the British and Dutch construction of aircraft noise indexes serves as a magnifying glass for studying both national and international efforts involving the formulation of standards. By also briefly touching on relevant examples from American and German approaches to aircraft noise, it becomes possible to unravel the traditions and ideals behind the proliferation of aircraft noise indexes in the 1960s. Let me first sketch the arrival of the American jets at Heathrow, and the response to what soon proved to be a booming business.

NEW SOUNDS IN THE SKY:
THE SCREAM AND WHINE OF TURBOJETS AND TURBOFANS

The American airliners started their first transatlantic jet service with the Boeing 707, adding the DC-8 somewhat later (Burns 1968: 208). The simultaneous introduction of economy class—to attract passengers for a doubled number of seats—and the halving of transatlantic flight time contributed to an impressive increase in air traffic. In 1958, for the first time in history, more people crossed the Atlantic Ocean by airplane than by boat. Transatlantic charter traffic by members of the International Air Transport Association (IATA) rose from 168,207 passengers in 1960 to 480,496 in 1965 (Bouwens and Dierikx 1996: 121–122). In 1961, Heathrow processed six million passengers, a figure that only four years later had risen to ten and a half million (Burns 1968: 208–209). Between 1958 and 1970, the aircraft industry delivered 3,757 turbojets worldwide.

Air travelers appreciated not only the reduced airfares and flight times, but also the relative tranquility of the jet cabins, a feature that was due in part to the new position of the engines. In contrast to this luxury, ironically, the noise emission of the jets dramatically increased compared to that of propeller airplanes. Whereas propellers made a "grumbling" sound, the new jets were "screaming" (Bouwens and Dierikx 1996: 124). As the historian of technology Erik M. Conway has explained, the turbojets were "larger, heavier, and more powerful than the last generation of piston-powered airliners." Further, jet engines "radiated most of their power at precisely the frequencies to which human hearing was most sensitive" (Conway 2003: 1–2). The acoustic energy produced by piston-engined planes had lower frequencies, and was less easily absorbed by the atmosphere. So, whereas the sound power level of Boeing's original 707 was 125 decibels, "its piston-engined rivals produced around 110 db" (Conway 2003: 6). Since the use of the decibel implied a logarithmic scale, this meant a striking increase in sound intensity.

This new noise soon aroused public concern, even if anxiety about sounds from the sky was not exactly new. As early as August 14, 1929, the *Times* had identified "the incessant roaring and booming of aeroplanes" as "the newest and most serious menace" to the peace of the public (Darby and Hamilton 1930: 39–40). The specific character of jet noise, though, was new *again*. Equally alarming was that jets needed longer runways, while residential areas increasingly encroached on airports as a consequence of urban growth (Kosten 1967: I/2; Robinson 1979: 122). At Heathrow, the number of complaints increased sharply from 1958 onward (Robinson 1964: 468). The Noise Abatement Law of 1960 was of no help to complainants, since it exempted aircraft noise from prosecution for causing nuisance. The aviation minister's decision, also in 1960, to allow night flights at Heathrow made things even worse, and not surprisingly, that same year the number of complaints about Heathrow peaked (Wilson 1963: 63).

The minister's ruling triggered fierce protests from the British Noise Abatement Society (NAS). The society had been founded the year before by the London businessman, John Connell. The summer of 1959 had been a hot one, with many windows open, and Connell was struck by the "many complaining letters" about noise in the newspapers. After he had discovered "the vagueness of the law in connection with noise," he sent a letter of inquiry to a few daily newspapers and received over a thousand replies, the "depth of feeling" of which inspired him to do something about it. Since the Noise Abatement League had perished in 1951 owing to a lack of funds, Connell considered raising a new Noise Abatement Society as a necessary first step. One of the responses to his newspaper letter came from E.J. Greaves, an "industrialist and financier with a London flat of 16 rooms complete with butler, manservant and four French maids." Connell took on board that he had found the ideal treasurer and decided to go along, without any financial help of the government (Connell 1960: 14–15; A Short 1968).

Aircraft noise became one of the key issues of the Noise Abatement Society. In the early morning of Saturday April 7, 1960, NAS president John Connell and his companions managed to haul their country's aviation minister out of bed. Carrying a noise meter with them, they "claimed the noise prevented them from sleeping and made life 'unbearable.'" Minister Duncan Sandys, dressed "in a silk dressing gown, blue pyjamas and red slippers," simply responded that noise near airports was unavoidable. "It is impossible—unless you close London Airport" (The Sunday 1960: 38). As the minister later underlined, "we must resign ourselves to the fact that noise and power go hand in hand" (Connell 1962: 9). Which, however, is noisier, "a Rolls Royce or a motor scooter," Connell rhetorically asked in the NAS journal *Quiet Please*. He felt that the world's greatest forces were "completely silent," notably "magnetism, gravity, electricity and, greatest of all, the power of thought" (Connell 1962: 9).

Despite his trust in the mind's silent power, Connell apparently considered it a strong move to take a noise meter along with him in trying to convince Sandys of the evil of aircraft noise. How to measure aircraft noise was far from self-evident, though. As discussed in chapter 4, acousticians had created an objective noise meter in the late 1920s, which succeeded the subjective noise meter. With a subjective noise meter, skilled operators had measured the loudness of a tone by changing the intensity of a reference tone until it was felt to be masked by the tone to be measured. In contrast, the objective noise meter consisted of a microphone, an amplifier, and an indicating meter, thus fully delegating the measurement of loudness to the meter.

By introducing the objective noise meter, experts in acoustics allied their field to the ideal of mechanical objectivity. As the historians of science Lorraine Daston and Peter Galison (1992) have shown, the moral of mechanical objectivity emerged in the late nineteenth century and reached its zenith in the 1920s. Mechanical objectivity originally aimed at eliminating the subjectivity of aesthetic judgment and scholarly dogmas from the registration of natural phenomena, and it ended up dismissing the human mind and body as trustworthy witnesses of phenomena in favor of machines. Creating an image of a perfect example of a species, for instance, transformed into the ideal of merely showing a variety of particular specimens of that natural kind—a moral imperative that was assisted and strengthened by photography. The shift from subjective to objective noise meters in acoustics clearly expressed this new ideal of mechanical objectivity. The subsequent change in the name of these meters, from "noise meters" to "sound level meters" did the very same. In the 1920s, acoustic experts increasingly defined noise as "unwanted sound," the "annoyance" of which could, at least in principle, be independent from a sound's intensity. Therefore, "noise meters" became "sound level meters."

Yet the first sound level meters could not measure the equivalent loudness of impulsive sounds (produced, for instance, by motor vehicles and aircraft engines) as correctly as continuous sounds (Davis 1938). Also troublesome was the measurement of complex sounds of more than one frequency. After World War II, Harvard professor S.S. Stevens introduced a method for calculating such complex sounds that broke with the notion of summing up "the loudnesses in the several octave bands into which the noise could be analyzed" (Stevens 1956: 807; Robinson 1957). The underlying idea was that the phenomenon of masking influenced the total loudness of complex sounds: it was "rather less than the sum of the loudness of its component parts" (Robinson 1960: 4) Soon afterward, the German acoustician Eberhard Zwicker would create an even more refined procedure that took account of asymmetries in masking (Schultz 1972: 18–19).

These were the kind of complexities anyone wishing to control noise had to deal with when turbo-jets entered the scene. In 1957, the director of the Port of New York Authority[3] so much feared the upcoming arrival of the turbo jets and the lawsuits from citizens he expected in their slipstream that he wanted to have a norm that "the city would force the airlines to abide by." Unlike the United Kingdom, the United States allowed for such lawsuits. The port

authority's director requested Leo Beranek of the American firm Bolt, Beranek
& Newman to come up with a criterion, making clear that he "wanted the
jets to be no more annoying to the surrounding communities than the piston-
engined aircraft already were" (Conway 2003: 7). By refusing landing rights
to airlines not meeting the equal annoyance criterion in the future, he aimed
to force the airlines to start using suppressors. In the early 1950s, British and
U.S. aircraft engineers had worked on the noise reduction of jets not primarily
because they expected public concern, but because the immense acoustic energy
of the jets produced a dangerous acoustic fatigue that damaged the aircraft.
Suppressors had been the result, but several U.S. airlines were opposed to them
because they were heavy and caused a loss of thrust (Conway 2003: 5–7). The
motivation behind the port authority's search for a standard was to get these
suppressors back on the agenda in order to protect the authority against citizens
who might want to sue it.

    According to Beranek and his associates, the application of Stevens's Mark
II loudness formula worked relatively well for propeller aircraft, yet it "underes-
timated" the "nuisance from jet noise." Therefore, they dismissed "the loudness
criterion" as inappropriate. They based their measurement scale, the PNdB (Perceived Noise level in decibels), on research in which observers judged sounds
"in terms of 'noisiness' rather than 'loudness.'" The result was a formula "strictly
analogous to Stevens's, but it differed in numerical detail, notably by giving
more weight to the high frequencies in the noise" (Schultz 1972: 31). It thus
reflected "the fact that people judge higher frequencies to be more annoying or
less acceptable than lower frequencies when factors such as 'meaning,' novelty,
adaptation, etc. are held constant" (Beranek et al. 1959: 23).

    Given the high frequency noise of the jets, meeting equal annoyance of
turbojet and piston-engined aircraft would mean that the jets had to be silenced.
Not surprisingly, Boeing was not amused, and "ran an experiment using its
own employees as subjects. It placed groups of employees in a house it owned,
subjected them to overflights of various types of aircraft, and recorded their
responses. To give the tests some credibility, it also invited Leo Beranek and

others to watch the tests and check the results. But unfortunately for Boeing, its employees had the same reaction to jets that [other] subjects had had" (Conway 2003: 8).

Consequently, the Port of New York Authority based its civil aircraft noise limit, the first of its kind, on PNdB: a maximum of 112 perceived decibels measured outdoors under the flight path for operations between 8 a.m. and 10 p.m. (Beranek et al. 1959: 29). This maximum had been the former average propeller-driven aircraft level (Conway 2003: 8–9). The norm implied the port authorities' original worry about the civil lawsuits to be expected after the introduction of the jets, and thus the U.S. legal context itself. The Composite Noise Rating (CNR) for Commercial Airports (1964), again designed by Bolt, Beranek & Newman, was motivated by a similar fear of legal actions from citizens. This rating—or aircraft noise index—was framed in terms of observable public response, ranging from the expectation of essentially "no complaints" at the lowest levels of noise, via potential "vigorous" complaints from individuals, to "Concerted group action might be expected" at the highest levels (Schulz 1972: 39).

The new maximum aircraft noise regulation at the New York airports did not immediately calm the public down. The U.S. jet airlines initially attempted to meet the norm by applying noise suppressors and using a new takeoff procedure—climbing as steeply as possible right after takeoff. Yet, since pilots considered this new takeoff procedure "a gross violation of their safety training," the airlines did not "force the issue." An "explosion of anti-noise activity" was the response (Conway 2003: 9).

Despite the protests against aircraft noise, it was not before the mid-1960s that the aircraft industry started research into the reduction of noise at its source: by silencing the engines themselves. In jets, a high-velocity stream of gases causes a turbine to rotate. The gases originate by burning fuel in a combustion area, a process that requires oxygen derived from atmospheric air and compressed in a compressor. The turbine, the compressor, and the jet itself are thus all sources of noise. The jet engine's "characteristic roar," however, is the result of the "violent mixing" of the exhaust gases with the surrounding air (Burns 1968: 212; Wilson 1963: 63). One solution was to create suppressors that redirected the flow of gases to make the exhaust process less chaotic. But the suppressors reduced the efficiency of the engines. Another solution was to reduce the velocity of the gas, while compensating for the loss of thrust by increasing the mass of air handled by employing a turbofan engine. The disadvantage of the turbofan was that it produced pure tones—the typical turbofan whine. Yet, by creating a higher bypass ratio in the turbofan engine and applying acoustic material, the aircraft industry was able to suppress "the tonal fan component of the frequency spectrum, in effect shifting the engine's noise signature toward the lower end of the spectrum" (Conway 2003: 16). Boeing even managed to get the level down to just 90 PNdB, much lower than the specified takeoff limit (104 PNdB) prescribed by the U.S. Federal Aviation Administration for new aircraft in 1966.

According to Conway, their success had again resulted from economics more than from environmental concerns, since the high-bypass turbofan engine "came with a dramatic increase in fuel efficiency" (Conway 2003: 17; Robinson 1979: 122). Yet, the high-bypass turbofan was also the result of subsidies the American Air Force provided to the aircraft industry in a 1964 design competition for a huge transport airplane for troops and heavy equipment. For this competition, Boeing designed its wide-body jumbo Boeing 747 with a high-bypass ratio turbofan jet engine (Bouwens and Dierikx 1996: 234). Those living near airports, however, were not going to profit from this invention anytime soon, since the life cycle of airplanes was 25 years, meaning that traditional jets would be around for years to come.

So, when the British government appointed a Committee on the Problem of Noise, chaired by Sir Alan Wilson in 1960, the jets were still roaring. Although Heathrow had adopted a maximum in PNdB close to the American one, and an even stricter night norm, the minister of aviation sought the cooperation of airlines instead of barring them from using the airport (Burns 1968: 214; Wilson 1963: 67–68). Before deciding on additional forms of aircraft noise control, however, the Wilson committee commissioned two studies: a social survey at Heathrow and a jury test at an Aircraft Show in Farnborough, led by

D.W. Robinson, one of the leading experts in acoustics at the National Physical Laboratory, London. On receiving the results of these studies, the Noise Committee would establish its aircraft noise index. If it was obvious that the New York maximum and the American Composite Noise Rating for commercial airports had been inspired by the American legal context, what were the origins of the British noise index, the NNI? Where, for one, did the idea of a jury test come from?

JURIES OF LISTENERS: THE BRITISH APPROACH

In 1931, the journal *Nature* published a paper by G.W.C. Kaye, head of the Physics Department at the National Physical Laboratory (NPL). The NPL, financed by government grants and fees for commercial testing, dealt with metrology—the standardization of physical constants and the calibration of measuring instruments—yet also gradually took up an elaborate research program (Magnello 2000). According to Kaye, a leading expert in British metrology, it was evident that the issue of noise measurement was "one of some complexity, involving not only physics but also physiology and psychology. Nevertheless, so far as the physical aspect goes, it is clearly desirable that there should be a consensus of opinion on the choice of a system of physical quantities. They should be preferably of an absolute character, so as to assist, inter alia: (a) in translating vague aural judgments and comparisons into facts and figures; (b) in elucidating the causes and characteristics of noises; (c) in comparing the results of different investigators; and (d) in setting up such arbitrary standards of noise as may be desired in the light of social, technical, or legal requirements" (Kaye 1931a: 257–258; see also Berry 1998).

In terms of recent literature on standardization, what Kaye was looking for was a technology of trust, coordination, regulation, and control (Porter 1995; Egyedi 1996; Schmidt and Welre 1998). A "system of physical quantities" had to eliminate variation of judgment, thereby delegating trust from the human ear to instruments and methods of measurement. Such a system would enable one to compare the results of investigators and thus to coordinate their work. And last but not least, it would help the worlds of politics and law to create standards so as to tackle the problem of noise through regulation and control.

That the measurement of noise was an issue, in the words of Kaye, "of some complexity" was quite an understatement. And not only measuring noise, as we have seen, proved a great challenge; the same difficulty applied to understanding annoyance. A few years before Kaye's paper appeared in *Nature,* the NPL, by order of the British government, had tried to find criteria for the "stridency" of motor horns, as mentioned in chapter 4. At first NPL physicists were quite confident about their ability to sort out the right from the wrong sounds. They set up an experiment in which they measured the loudness, acoustical pressures, and the average waveform of the sounds produced (using an instrument called the oscillograph) by different horns. "It was thought that simple inspection of the oscillograph records might provide an indication of stridency, but no such correlation was observed." In contrast, the researchers found out that "sheer loudness was the primary consideration in accounting for objectionableness."[4] (Figure 7.1.)

They still thought it feasible to find a relation between, for instance, "strong notes unrelated to the general harmonic series" and "stridency."[5] That in 1937 the NPL Executive Council stressed that the lab's research should be "of benefit to the country" may have helped to stimulate their search (Magnello 2000: 97). Moreover, the Ministry of Transport kept asking for it. So the NPL physicists extended, in the very same year, their motor horn studies by asking a jury of "some 200 listeners of both sexes, and of ages between 16 and 60" to evaluate the sounds of a large number of horns, including "bulb horns, electric buzzer horns, wind driven horns, 'klaxon' type horns and sirens."[6] Yet again, "while no single measurable characteristic of a horn serves entirely to determine its objectionableness, the loudness of the sound is of major significance."[7]

Equally important as these results, however, was the fact that the NPL transferred the jury tests they employed for studying the noise of horns to motor vehicle noise—which began in the late 1950s—and then to aircraft noise in the 1960s. In a 1959 motor vehicle noise study, for instance, a jury of nineteen listeners estimated the noisiness of 225 randomly passing vehicles on the A23

Figure 7.1 The horn noise tests at the National Physical Laboratory, London, 1929. Courtesy National Physical Laboratory, London. Courtesy of © Crown Copyright 1929.

London–Brighton highway (Robinson et al. 1961; Mills and Robinson 1961). (Figure 7.2.) By doing so, the NPL moved not only the idea of outdoor jury tests from the noise of horns to that of traffic, but also the associated apparatus. For instance, a mobile laboratory, the first of its kind in the country (Pyatt 1983: 189; Copeland 1948), enabled the measurement of noise outdoors. A calibrated amplifier and a high-fidelity tape recorder were connected to a microphone outside. Yet it also "effectively screened" the listeners "from view of upcoming drivers" and thus helped the researchers cope with the peculiarities of field research (Robinson et al. 1961: 2).

According to Robinson, such jury tests offered a perfect "middle way" between laboratory studies and social surveys, between preciseness and validity, and "some compromise between scientific rigour and realism."[8] Outside, listeners had no trouble rating actual motor vehicle noise, he claimed, but regarding

Figure 7.2 The judgment of traffic noise by a jury, tests organized by the National Physical Laboratory, London, 1950s. Courtesy National Physical Laboratory, London. Courtesy of © Crown Copyright 1950s.

noises recorded and played back in a sound-absorbent room, "our judges did not seem to be able to make up their minds at all."[9] On similar grounds he distrusted indirect noise assessment methods based on observation of behavior, such as the effect of noise on job performance, because one might as well test something "quite different from annoyance."[10] In case the subject of study involved "a psychological complex attribute such as 'acceptability,'" he argued, "it is useful—if not indispensable—to employ real life sounds. Assembling a jury guarantees that the acoustical stimuli are the same for all, it facilitates the objective measurements and permits the use of a variety of psychophysical test procedures into which consistency checks can be built" (Robinson 1967: 397).

Evidently, realism in the sounds evaluated was highly important to Robinson, as was the ear-witness aspect of the experiment. To him, "a jury of listeners" represented "the average citizen." The test results, though, still needed interpretation. One had to decide, for instance, whether one would satisfy the "average man" or *everyone* on the jury," as Robinson explained at an international NPL conference on noise control in 1961 (The Control 1962: 15). Jury testing, in other words, did not free the world of politics from deciding on how many people were "allowed" to be annoyed. In addition, polls could be employed so as to gauge "the significance of aircraft noise nuisance ... in relation to other environmental factors," thus providing "the final value judgement upon which administrative action depends." Still, a poll was an "expensive" tool and a "rather blunt weapon." Because one could not, as in an experiment, fully control the physical stimulus, the "numerical significance of the results" was limited and could be used only "within the limits of the prevailing conditions" (Robinson 1967: 397–398; Robinson 1973). Therefore, jury testing continued to be the best way to start.

At the very same time, Robinson's crusade for jury testing in aircraft noise research was in line with the NPL motor horn experiments of the 1920s and 1930s and the motor vehicle studies of the late 1950s and early 1960s. As such, his efforts also illustrate the consequences of a dominant paradigm, understood by Thomas Kuhn (1970/1962) as a set of problem definitions, procedures, and knowledge claims, that, embedded in institutions, infrastructures, and artifacts, constrain the choices of the researchers involved. It is important to note, however, that I can substantiate such streamlining effects only in the acoustic research that the NPL carried out by direct order of the government and on issues that concerned the judgment of the "unpleasantness" of noise. The acoustics research at NPL had originally started with testing tuning forks, measuring underwater noise, and studying radio sound transmission (Magnello 2000). It proceeded with examining the methods and instruments of sound measurement, the material reduction of aircraft, road and rail noise, and sound transmission and absorption in buildings. It also involved fundamental laboratory research on equal loudness contours and the variation in loudness with the direction from which a sound arrives relative to an observer's head.[11] Yet, the NPL had found a distinct style of studying those kinds of noise that the government aimed to ban or to limit through legislative intervention. The Farnborough aircraft noise studies clearly built on the research culture that had been developed in the motor horn and vehicle noise studies.

The large significance of this tradition of research also becomes clear from the tension between Robinson's defense of NPL's jury testing vis-à-vis other approaches and his disappointment about the proliferation of national aircraft noise standards. As he claimed in 1967, the aircraft noise situation could not wait for perfection, as it was "easier to modify and improve in a climate of initial agreement than to come to agreement later when divergent—and perhaps irreconcilable—practices may have become embedded in national regulations and legislation" (Robinson 1967: 396). This conviction, however, did not stop him from advocating his belief in jury testing, thus revealing how deeply rooted the differences in expert opinion on assessing aircraft noise could be. It also illustrates that Robinson was highly aware of how standards are practically and materially entrenched. Historians and sociologists of science have shown that standards do not work without the routine skills, hands-on instruction, disciplined bodies, material setups, and values connected to the laboratories and research institutes from which they stem. Exactly because standards have never been neutral but embody local interests and time-bound morals, they are in dire need of many companions to stay robust "beyond there immediate surroundings" (Schaffer 1999: 476; Schaffer 2000). Or in the idiom of Bruno Latour: that standards "act at a distance" is never self-evident and requires the mobilization of "large numbers of allies" (Latour 1987: 228, 240). Once established in larger areas, the standards are often so intricately connected to the networks that enable them to function that they are far from easily replaced by new ones, as Robinson already seemed to have been aware of before the publications of Latour, Schaffer, and others (Jasanoff and Wynne 1998).

In the Farnborough study executed at the request of the Wilson committee, sixty subjects of varying age and sex judged the sounds of real aircraft flying overhead on three afternoons during the 1961 Aircraft Show at Farnborough.

Figure 7.3 The judgment of aircraft noise by a jury, tests organized by the National Physical Laboratory (London) during the Farnborough Air Show, 1961. Courtesy National Physical Laboratory, London. Courtesy of © Crown Copyright 1961.

(Figure 7.3.) A comparison of the aircraft and motor vehicle jury tests showed that the aircraft and motor vehicles were "both judged 'quiet' for the same sound level at the ear, about 66 dB(A). . . . At higher levels, subjects are more tolerant of aircraft than of motor vehicles, the difference amounting to some 20 dB(A) for the criterion 'very noisy'" (Robinson et al. 1963: 335). Somehow, the researchers suggested, the subjects judged all the noises within a familiar range between their concept of "quiet" and "very noisy," and they changed their scale-factor for different types of noises (Robinson et al. 1963: 331). This meant that the identification of the nature of the source of the sound was related to the acceptability of particular sounds, apart from criteria already known, such as the sound level involved.

The Wilson committee digested these and other results and published its report in 1963. Its members included a professor of psychological medicine, a professor of aeronautical engineering, a medical officer of health, an employee of a development corporation, a town clerk, a housewife, and six representatives from the industry, such as Rolls-Royce and a manufacturer of both automobiles and aircraft engines. Of the nine adjudicants, seven came from public authorities related to medicine, housing, transport, labor, and aircraft, and two came from research institutions. One of them was Robinson. His influence became clear in the explicit support for the juries of listeners' assessments, since the committee felt "that in these experiments the nearest practicable approach has been made to a measurement of the opinion of the man in the street" (Wilson 1963: 47). The voice of the industry was prevalent as well. The interests of airlines and the costs related to reducing the noise of aircraft engines were often emphasized (Wilson 1963: 66, 68, 70, 71, 76–77).

As mentioned, the British Noise and Number Index became partly based on a social survey conducted, again, at the request of the Wilson committee. If the committee admitted that aircraft noise was "an international problem," it considered the circumstances at Heathrow so different that "it proved impossible to make detailed comparisons between the problems at Heathrow and at other airports" (Wilson 1963: 69). Hinting at the fact that the areas around Heathrow were densely populated, they called for a "survey of subjective reactions of people living near the Airport" (Wilson 1963: 74). Moreover, as even Robinson had admitted, the poll could provide information on the disturbance of aircraft noise in comparison to other inconveniences, which would enable authorities to decide on administrative action. At the very same time, however, the committee used the results of Robinson's Farnborough study to "calibrate" the outcome of the social survey, apparently following his trust in the jury tests.

The survey was carried out by the Central Office of Information and led by A.C. McKennell. In the summer of 1961, some 1,909 adults living within ten miles of Heathrow were interviewed. In addition to questions about aircraft noise, they were asked questions about the general living conditions in the area. Such questions were included because the "seriousness of the aircraft noise nuisance might reasonably be doubted . . . if this did not feature at all or only minimally among those aspects of their environment reported, spontaneously, as disturbing by residents in the region of the airport" (McKennell 1963: 3–1). The residents were also asked whether or not aircraft noise ever woke them up, interfered with listening to TV or radio, made the house vibrate or shake, interfered with conversation, or bothered them in any other way. Their responses were rated between zero and six, with zero indicating "no annoyance." So few of the adults interviewed gave answers that fell within the maximum of six, that five and six were combined into one "extreme" category, while three stood for "moderately annoyed" and four for "very much annoyed" (Burns 1968: 221; Wilson 1963: 205–206). In the meantime, the Ministry of Aviation recorded one hundred successive aircraft at eighty-five points distributed over the area under investigation. Of all the information gathered, only the average peak noise level in PNdB and the number of aircraft heard during the day were claimed to be relevant.

When average annoyance scores of groups of people were compared with averages of numbers of aircraft heard (within particular noise ranges), the results displayed an orderly relation (Wilson 1963: 207). McKennell noted, however, that "very considerable variations in individual susceptibility to the aircraft noise" were found. Even in the "quietest locations" about 10 percent of the persons surveyed felt seriously disturbed by noise, whereas in the noisiest areas 32 percent were not seriously bothered, a group he labeled the "hard core of imperturbables" (McKennell 1963: 2–3). Given the large variations in individual susceptibility to aircraft noise, the Wilson committee claimed that "only very great changes in its characteristics can have a substantial overall effect" (Wilson 1963: 217). The committee therefore decided to focus on the areas in which a large majority of residents reported intense annoyance. (Figure 7.4.)

What the committee was searching for was a formula that would show the options for intervention if one aimed to keep annoyance at a particular level. If the average noise level of the aircraft heard would increase, how much reduction of the number of aircraft heard would then be necessary to keep the annoyance constant, and vice versa? This "trade-off" between noise and numbers was

(i) Noise level in PNdB	(ii) Average number of aircraft per day	(iii) Annoyance score						(iv) Average annoyance score	(v) Number people in in stratum
		0	1	2	3	4	5		
84–90 ...	5·75	230	128	113	5	5	31	1·1	512
	22·5	45	33	26	17	12	22	1·9	155
	81	5	7	2	7	10	7	2·8	38
91–96 ...	5·75	51	41	28	17	11	10	1·5	158
	22·5	90	64	55	45	35	32	1·9	321
	81	18	15	13	23	18	23	2·7	110
97–102 ...	5·75	2	1	—	3	1	—	2	7
	22·5	13	9	20	16	11	13	2·5	82
	81	20	22	38	26	30	64	3·1	200
103–108 ...	5·75	—	—	—	—	—	—	—	—
	22·5	1	—	1	5	2	2	3·2	11
	81	11	7	17	16	19	67	3·6	137

Figure 7.4   The number of people with various annoyance ratings classified by noise level and number of aircraft per day. From A.H. Wilson, *Noise: Final Report, Presented to Parliament by the Lord President of the Council and Minister for Science by Command of Her Majesty, July 1963* (London: Her Majesty's Stationery Office, 1963), p. 207. Courtesy Office of Public Sector Information (OPSI), PSI-License C2007001229.

this: "a fourfold increase in the number of aircraft heard is very approximately equivalent to a rise in average peak noise level of 9 PNdB" (Wilson 1963: 75). On this basis, "an estimate of the total noise exposure causing annoyance" could be obtained in terms of "the sum of the average peak noisiness in PNdB and 15 times the logarithm of the number of aircraft heard" (McKennell 1963: 5–3). When combined with what the committee determined as the level of zero annoyance for one aircraft, about 80 PNdB, the Noise and Number Index (NNI) became

$$NNI = \bar{L}_{PNmax} + 15 \log_{10} N - 80$$

where L_{PNmax} is the "average (taken logarithmically) of the maximum perceived noise levels attained during the passage of successive aircraft" and N is "the number of aircraft heard in the defined daytime period."[12] For an aircraft to be "taken into account for evaluation of NNI, the maximum perceived noise level at the position in question" had to exceed 80 PNdB (Noise Units 1975: 3).

Another result the study found was that at 55 NNI, over "three quarters of people suffered interference with conversation" (Burns 1968: 228). The Wilson committee combined this with information from the McKennell report about the indirect effects aircraft noise, such as the relation between PNdB and the percentage of people wishing to move in order to get away from it compared to those wanting to move for other reasons. It showed that at levels above 50 NNI, aircraft noise began to predominate over other reasons for people wishing to move. The committee therefore claimed that "aircraft noise becomes excessive by any reasonable standards, when the noise exposure allows values of 50 to 60 NNI" (Wilson 1963: 75). As for the maximum NNI levels at night, the committee formulated only tentative conclusions. The Wilson committee's last and decisive step was to relate the results of the survey with those of the Farnborough tests. This made clear that the Farnborough figures were "smoothly related to the NNI" (Wilson 1963: 209). The committee concluded that the 50–60 NNI levels corresponded "with average judgments around 'very annoying' on the scale of intrusiveness used in the Farnborough experiment" (Wilson 1963: 75).

Measured in NNI, the Wilson Committee mapped the contours of aircraft noise heard in the vicinity of Heathrow Airport, which showed its contours in intervals of 5 from 50, to 70 NNI. In doing so, the committee linked the idea of zoning to particular levels of noise impact, an intervention plan that coupled a strategy of noise abatement that had emerged in the context of industrial noise to a solution that had been designed to control traffic noise. The map played a role in the British government's decision in 1965 to provide people living in the area near Heathrow, which was expected to have an NNI of 55 by 1970, with grants covering half the cost (up to a specified maximum amount) of soundproofing up to three rooms (Bürck et al. 1965: 109). Moreover, the NNI became the basis of land use policies concerning the development of areas around airports (Starkie and Johnson 1975: 18–19). Additional measures taken at Heathrow included the employment of Minimum Noise Routings and a Preferential Runway System (Starkie and Johnson 1975: 103). Furthermore, the Civil Aviation Act of 1968 created the right to prohibit aircraft from taking off or landing in the U.K. in cases in which they did not comply with noise certification regulations (The Law 1969: 33–34). Two years later, in 1970, the U.K. passed the Air Navigation Order, which created a national noise certification system of civil aircraft, ratified by the International Civil Aviation Organisation (ICAO) in 1971 (Starkie and Johnson 1975: 13; Kerse 1975: 83).[13] The Control of Pollution Act of 1974, however, did not yet apply to aircraft noise (Garner 1975: 2).

In 1971, a second, larger survey of aircraft noise annoyance near Heathrow created serious doubts about the appropriateness of the NNI. The survey reminded the examiners that "certain cells in the loudness-by-number-matrix" used in the Wilson report had been "empty or very poorly represented in the sample." They therefore wanted to check the "'trade-off' between loudness of aircraft and the number heard" (Second 1971: 2). It did not hold: "Whereas for each level of loudness the degree of annoyance has stayed the same between 1961 and 1967, for each level of number of aircraft, the degree of annoyance has fallen" (Second 1971: 8). Consequently, the minister of trade asked the Noise Advisory Council, established in 1970, to advise him on what to do about the NNI. Remarkably, it advised the minister *not* to revise it, suggesting that the smaller degree of annoyance found in 1967 compared to 1961 might be attributable to the self-selection of the people moving in and out of the area, habituation, or even sheer resignation and apathy (Aircraft Noise 1972: 3). Moreover, the relaxation of the NNI would become "perpetuated in bricks and mortar," which would not be easy to reverse in case of a later revocation, and confusion would probably arise between the old and new NNI. Even worse, a modification of the NNI might "bring into disrepute the entire basis of such methods for quantifying nuisance from aircraft noise" (Aircraft Noise 1972: 5).

In sum, three kinds of constraints kept the British focused on a national noise standard, despite their awareness of the internationalism of the aircraft noise problem. The first was the paradigmatic research tradition of jury testing at NPL, tunneling the search for an aircraft noise index in line with research into the types of noise that had been listed on previous political agendas. The second was the conviction that the situation at Heathrow was unique. The third, finally, was the fear of the public's rejection of any noise standard as soon as one would revise—and thus raise doubts about—the one at hand.

The influence of research traditions on the establishment of aircraft noise indexes is also evident in the making of the German *Störindex Q*. The *Störindex* was first proposed in 1965 in a report prepared by order of the German Ministry of Public Health. The committee charged with writing the report stressed, as Robinson had, that only the international regulation of aircraft noise could offer a satisfying long-term solution to the aircraft noise problem because of the interwoveness of air traffic (Bürck et al. 1965: 88). Moreover, the committee members extensively reviewed aircraft noise indexes used in other Western countries and even visited the United Kingdom and the Netherlands (Bürck et al. 1965: 97). They did not want to use the NNI, however, because the British had not taken the duration of noise into account in devising it (Bürck et al. 1965: 129, 132–133).

The committee's evidence for the significance of the duration of noise was based largely on studies carried out within a long-standing German tradition of research on the physiological effects of industrial noise, such as hardness of hearing and irritation of the nervous system. Among the works referred to by the committee were the laboratory studies of two of the five members of the German committee: the acoustic engineer Franz Joseph Meister and W. Bürck.

The first study had demonstrated a relation between the duration of noise and hearing fatigue. The second study had established a link between the duration of noise and disturbance in work (Bürck et al. 1965: 130–131). The other studies discussed came from Gunther Lehmann. Lehmann was a physician, the director of the Max Planck Institute of Industrial Medicine, and the president of the German Working Group for Noise Abatement. Remarkably, he defined noise not only as annoying sound, but also as sound that could harm humans (Lehmann 1961: 7).

Lehmann claimed, for instance, that the peripheral blood vessels of the skin narrowed under the influence of noise, which he determined would occur at a level of above 65 phon during the day and 50 phon during sleep. The narrowing of the blood vessels caused an increase of the diastolic blood pressure, which, in turn, affected the blood's circulation. The effect was independent of the pitch of the noise or a person's subjective response or habituation to noise. Yet the time period needed for a person's pulse to return to within normal levels was strongly dependent on the duration of noise (Bürck et al. 1965: 25–26, 131; Lehmann 1961: 11–14). When subjected to continuous or recurrent noise, the increased tension of the body decreased blood circulation. In such a state, the body became vulnerable to other problems, such as (slight) disturbances in metabolism and heart rhythm, a pale skin, and various psychological problems. The situation at airports was comparable to workplaces, Lehmann asserted, where fourteen machines were started, one shortly after the other, the after effects of which exceeded the duration of the noise (Lehmann 1961: 21). According to the German committee, in terms of the annoyance created, a doubling of the duration of noise was equal to an increase of four dB ($\alpha=3/40$) (Bürck et al. 1965: 132, 140, 166).

Apart from duration of noise, the *Störindex* Q also considered the deviation of departing planes from their ideal runway. This resulted in the formula

$$\bar{Q}(x,y) = 1/\alpha \, \log_{10}\{1/T \int_{-b(x)k}^{b(x)} \Sigma \, 10^{\alpha Q}{}_{k.} \tau_{k.} n_{k} \, d\eta\}$$

where $\bar{Q}(x,y)$ is the noise impact at a specific spatial point, α is the equivalence parameter ($\alpha=3/40$), T is time, b(x), etc., is the deviation from runway, k is the

Figure 7.5 The measurement of noise of a Pan American airplane, [Germany], 1960s. From Franz Joseph Meister, *Geräuschmessungen an Verkehrsflugzeugen und ihre hörpsychologische Bewertung* (Cologne and Opladen: Westdeutscher Verlag, Arbeitsgemeinschaft für Forschung des Landes Nordrhein-Westfalen, 1961), 94:135.

class of airplane, Q_k is the maximum level of noise produced by an airplane of class k, τ is the duration of the noise, n is the number of airplanes heard, and η is the point above which the airplane flew (Bürck et al. 1965: 142–147). (Figure 7.5.) The formula was meant to reflect the relatively strong level of annoyance created by flights made at night through a doubling of n in the evening (between 8 p.m. and 11 p.m.) and a quadrupling of n at night (between 11 p.m. and 6 a.m.) (Bürck et al. 1965: 147, 165).

The German aircraft noise index, then, like the British one, sprang from a particular research tradition with which the experts involved were acquainted, resulting in yet another national aircraft noise index. The maximum of allowable noise that was eventually adopted was, as the German committee emphasized, arbitrary, given the variety of individual responses to noise. The committee nevertheless suggested, like the British, a zoning system. In Zone I (Q>82), residential housing should be prohibited. For Zone II (Q=77–82), one should not plan residential housing and should enforce soundproofing for the remaining homes. Even in Zone III (Q=72–77), residential housing should not be recommended and public schemes only granted in case of soundproofing. In the last zone, Zone IV (Q<72), residential housing could be planned, although the area would not be suitable for hospitals, schools, nursing homes, health recruitment homes, churches, and scientific institutes: the islands of silence we have become acquainted with in the previous chapters. Although the details of these norms did not fully survive the decisions subsequently taken by the German government, the index itself did, but only in Germany. The Dutch decisively departed from both the German and the British approach. Why?

An Odyssey Ending at Home: the Dutch Kosten-Unit

Among the many people attending the 1961 NPL conference on noise control were a few Dutch. One of them was C.W. Kosten. At first sight, he was the Dutch "Robinson." Like Robinson, Kosten earned his living primarily as a scientist. He was professor of technical physics at Delft University of Technology (de Lange and Janssen 1973). He collaborated closely with the Division of Applied Physics at the Dutch Institute for Applied Scientific Research (TNO), which, like the NPL, was an institute for applied research in physics. Yet, unlike the NPL, it was independent from the Dutch government. As with Robinson, Kosten had been asked to advise the government on aircraft noise and also strongly favored an international approach, as evidenced by his active membership in several working groups within the Technical Committee 43 on Acoustics, a committee within the International Standardization Organization (ISO) that was responsible for examining issues related to the measurement and assessment of aircraft noise (Kosten 1967: XII/4). He proposed yet another index, which was partially accepted by the Dutch government.

Kosten got his opportunity to influence Dutch politics through the Netherlands' Advisory Committee on Aircraft Noise Nuisance. This committee was established in 1961 with Kosten as its chair. It was asked to advise the government on the maximum levels of aircraft noise that would be acceptable and permissible without unnecessary interference in aircraft activities.[14] The clash between the interests of aviation and man's need for rest, the Dutch minister in charge claimed, needed the helping hand of modern techno-science.[15] Unlike the Wilson committee in the U.K., Kosten's committee had no representatives from the aircraft industry among its members, a measure intended as a warrant of its impartiality.[16] One of Kosten's colleagues at Delft University, the engineer G.J. Kleinhoonte van Os, was appointed committee secretary. The other members came from the Civil Aviation Authority and the ministries of Defense, Social Affairs and Public Health, and Housing and Planning.

It would take the committee six years to publish its final report, which appeared in 1967. Kosten advocated a comprehensive approach that focused on long-term, acoustics-based planning and surveillance of residential, recreational, and industrial zones instead of on piecemeal problemsolving.[17] Nevertheless, in the end Kosten was forced to take a piecemeal approach because of the way the minister continued to ask the committee to produce report after report on plans for a fourth runway-in-progress, nicknamed the Roaring Runway (*Bulderbaan*), and a fifth one, both of which committee members were very critical of. The minister's requests for ad-hoc advice was not the only reason for the delay of the final report, however. Like the Wilson committee, the Kosten committee considered it absolutely necessary to conduct a survey of residents who lived in the area near Schiphol.[18] Because of the large interests at stake, it refused to rely blindly on foreign research results and aircraft noise standards.[19] The result was a time-consuming survey of 1,000 residents, carried out between 1963 and 1964, and led by the psychologist C. Bitter who was affiliated with the Institute of Preventive Health of TNO (TNO-PG).

The third factor that slowed down the committee's work requires more elaborate discussion. It involved tensions between the supranational approach the committee felt was absolutely necessary and the various dead ends it encountered when trying to initiate such an approach. The committee believed that the Netherlands had to address the issue from a supranational level because a small country could only afford to exclude noisy aircraft from its airports if its neighbors would do the same. Planning future airports, especially those that would be placed near borders, would be impossible without international agreements. Even airlines, knowing that one day their aircraft potentially would be refused landing rights at particular airports, wanted to know where they stood and needed international guidelines.[20]

It was hardly obvious, though, which international body would be the most appropriate to take the lead. The Organization for Economic Cooperation and Development (OECD)'s Directorate for Scientific Affairs had established a commission aimed at furnishing data to aid member states in making decisions about aircraft noise control. But it met only once a year. An alternative was the ISO's Technical Committee 43 on Acoustics. The ISO, however, was allowed to make recommendations only on the definition and measurement of noise, not on what the maximum levels of permissible noise should be, and it operated even more slowly than OECD's Directorate for Scientific Affairs.

The Kosten committee felt that a far better candidate was the International Civil Aviation Organization (ICAO). The ICAO had recently asked the ISO's Technical Committee 43 on Acoustics to study aircraft noise nuisance. Yet again, it could take years before its results would allow for specific action. And Robinson had claimed, as a note on one of the Kosten committee's documents revealed, that "ICAO is not appropriate!"[21] Robinson may have known already what Kosten would soon find out. In March 1964, Kosten visited the Dutch representatives at ICAO in Montreal to discuss the need for international action. Kosten complained that it had failed to send a representative to the meetings of the ISO's Technical Committee 43 on Acoustics, presumably for financial reasons. One of the Dutch representatives suggested, however, that the real reason was that ICAO feared the consequences of aircraft noise regulation for the competition between air and other forms of transportation. The only way to get the ICAO moving would be an initiative taken by another international body, preferably the World Health Organization. To Kosten this seemed like a good idea: "Whatever Uncle Doctor claims, is true by definition!"[22]

From the perspective of the ICAO, it had been the ISO that had made a mess of the process of coming to an international agreement. The Kosten committee's archives include an intriguing report from ICAO-representative

A. Spooner, which described his impressions of a joint meeting held in Prague in 1966 of two of the Technical Committee 43 on Acoustics' working groups, one that was concerned with the measurement of aircraft noise and another that worked on its assessment.[23] The discussion centered on whether one should take dB(A) or PNdB as the starting point for describing aircraft noise around airports. As we have seen, the United States and the United Kingdom had already begun using PNdB. The ISO had implicitly accepted PNdB by a postal vote prior to the Prague meeting. In Prague, however, advocates of dB(A) took the chance to reopen the issue.[24] The result was a "long" and, as Spooner repeatedly pointed out, "difficult" discussion with representatives of the United Kingdom, the United States, Denmark, and the OECD as PNdB proponents, and the representatives of Germany, the Netherlands—probably Kosten and Kleinhoonte van Os—Austria, Switzerland, and Czechoslovakia favoring dB(A).[25]

Concerning the conflict itself, the dB(A), as Spooner summarized it, had the "virtue of simplicity" because one could read it from a readily obtainable sound level meter. In contrast, the PNdB was neither a straightforward unit, nor easy to calculate because it required "the use of a computer" as it was deduced from an analysis of recorded noise. But PNdB comprised "a more sophisticated 'weighing' of the sound level values" that could account for the high pitched noise of jets, and was therefore regarded as reflecting more clearly "the public's annoyance with the noise produced by jet aircraft in use today."[26] The heart of the debate, however, was about the specific goals for measuring aircraft noise to begin with, and thus, again, with the legal context in which policies had to be implemented. This becomes clear from Spooner's description of the conflict's closure. "Eventually," he explained, "it was agreed upon that monitoring in PNdBs was necessary principally in the case of aircraft operations where the purpose was to obtain evidence that could be used against airlines or that could substantiate claims made by members of the public. For these and similar purposes it was regarded as necessary to use the unit PNdB because of its high degree of precision."[27]

Because of the risk of civil lawsuits against aircraft, the United States certainly needed a unit appropriate for measuring public nuisance. The British representatives, in turn, had to reckon with the fact that their Noise and

Number Index and proposals for housing insulation grants were already based on PNdB. At the same time, the working groups "also recognized that large-scale monitoring of noise levels over a long period of time should be accomplished, indeed could only be economically accomplished, in dB(A) units. In this case, the purpose was to determine the overall noise environment by the use of sound level meters so that land zoning, town planning, surveillance and the like could be carried out."[28] As we have seen, zoning with surveillance was exactly what Kosten had in mind for his Dutch noise policy. The compromise reached under the joint chairmanship of the American and Danish representatives was that, for the time being, PNdB would be part of the draft recommendation for measuring the noise of a single airplane,[29] whereas the dB(A) advocates would still get a chance to have their draft recommendations discussed later on.

Spooner estimated that this was "way beyond the scope" of the ISO working groups' agendas. The International Air Transport Association also believed "that ISO might be going beyond its legitimate interest in the matter."[30] Clearly, much was concurrently at stake. Kosten preferred the ICAO to take the lead, yet it hardly seemed fond of aircraft noise restrictions. It feared a weakening of the market position of aviation vis-à-vis other kinds of transportation. Therefore, it did not allow the ISO to take the initiative in establishing them, at least not beyond defining the units and measurement of noise. Apparently, the ICAO had the International Air Transport Association on its side. The OECD found itself amid the parties fighting over dB(A) and PNdB, while the World Health Organization, mentioned as a potential arbiter, had not entered the scene yet. Moreover, the choice of a specific aircraft noise unit was not just a technical matter. The battle over dB(A) and PNdB reflected disagreement about which national policies would be allowed to influence future international policies.

A year after the ISO meeting in Prague, the Kosten committee published its final report, repeating at length its arguments for an international approach, but also arguing that the Dutch government should no longer wait for an international agreement. It was time to adopt a specific aircraft noise index and revise the formula later on if necessary (Kosten 1967: XII/6, II/16). The committee claimed that a new "Dutch" method was absolutely necessary, dismissing the

U.S. index as "needlessly inaccurate," the German one as "too thorough to be feasible," and the English method as "illogical" (Kosten 1967: II/8).

What, exactly, was wrong with the other indexes? Some, such as the NNI, showed a fairly good correlation with the results of the Dutch survey (Kosten 1967: III/9). Yet the NNI, like the American index, used PNdB and was thus not open to the direct control preferred by Kosten *cum suis* (Kosten 1967: II/14–15, III/10). Nor did the Kosten committee believe that the methods that took the duration of noise into account, such as the "thorough" German one, were practical because duration was not directly measurable (Kosten 1967: II/15). Worse, however, was the British choice to consider aircraft that produced less than 80 PNdB as not contributing to the noise to be calculated. The committee argued that one should count all aircraft creating 70 PNdB, which translated into 56 dB(A),[31] or more.[32] And while the NNI had no night penalty, the Germans calculated night flights in such a strange way that changing flights from day to night could, under particular circumstances, result in a lower noise impact.[33] What's more, the Germans had incorrectly taken industrial noise—with levels much lower than those of aircraft noise—as the basis for their unit (Kosten 1967: III/5).

These considerations as well as the annoyance measured by means of the survey resulted in a formula in which the noise impact B in a particular area was calculated on the basis of the number of passing aircraft per year (n), the maximum noise (L) in dB(A) created by each of these planes, and the moment of passing by

$$B = 20 \log \Sigma \ (n.10^{L/15}) - 157$$

In this aircraft noise index, the night factor came to be expressed as a correction of (n), which was given a value of 1 during the day and more than 1 in proportion with the increased nuisance caused as a consequence of aircraft passing by in the evening or at night.

As important as the standard itself was the norm associated with it. According to the committee, the percentage of people severely hindered by aircraft noise should not exceed one third of the population. The corresponding number of B, which later came to be expressed in Kosten units (*Kosten-eenheden*) or Ke, was 45. At B=45 (or Ke=45), the average relative annoyance score among the population was 45 percent (Kosten 1967: III/12). At this average, the survey revealed, 27 percent of the population reported frequent interference with communication, 33 percent reported sleep disturbance, and 66 percent reported occasional feelings of fear (Kosten 1967: III/15). To the committee, these findings meant that a third of the people involved "suffered serious annoyance" in terms of sleep and communication, while the figures on fear suggested that many failed to adapt to aircraft noise (Kosten 1967: III/16–17).

With respect to zones around airports with a noise impact of 45 Ke or higher, the committee recommended forbidding the planning of housing, as well as that of buildings such as offices, laboratories, nursing homes, hospitals, schools, and churches (Kosten 1967: VIII/3–4). To prevent situations in which the maximum would be exceeded, it was necessary to reduce aircraft noise itself through a certification policy, create distance between residential areas and airports, cut down on overhead aircraft, design preferential runway systems and noise-reducing flight procedures, and keep the public informed. In case of existing airports, and in zones with 45 Ke or higher, residential housing and other vulnerable buildings gradually had to be demolished. For unlike the Wilson committee, the Kosten committee was not enthusiastic about soundproofing housing with double-glazing: it considered a permanently closed home unfit to live in (Kosten 1967: II/10, VIII/2).

Although Kosten's committee cautiously sought to legitimate its 45 Ke boundary, its choice seemed rather "arbitrary," as historians of aviation have claimed (Bouwens and Dierikx 1996: 274). Documents, hitherto neglected by historians, from the archives of the institute that executed the survey—which closely resembled the design of the British one but was, unlike McKennell's, classified as "secret" and never made public—show why. On Christmas Eve 1964, Kleinhoonte van Os prudently asked Bitter, coordinator of the survey at the Institute of Preventive Health of TNO, to make clear at which score the noise became impermissible from a psychological point of view.[34] Apparently,

Bitter did not respond, since Kosten tried again at the end of April.[35] A month later, an elaborate letter from both Kosten and van Os followed in which they explained to Bitter that they could not satisfy the minister by just mentioning a particular annoyance score. "We *have* to explain what an area with a score of 1, 2, 3, 4, 5, 6 means, or, in other words, what the situation will be."[36] They compared their predicament with that of a statistician who, despite his weighing and measuring countless individual people, was still unable to say which weight was best at what length. "The physician, however, *is* capable of judging; at times he has to be. Regardless of whether he makes up the story or obtains one in a different way, he has to speak out. Probably he will found his judgment on statistical data, his experience and by feeling. Although it is clear that his opinion may have to be recalled on the basis of new data and despite the difficulties involved, a decision has to be made. With aircraft noise impact it is similar. We ask you to judge how bad the situation is in the areas of the survey.... The anamnesis has now been completed, and it is your task to do the diagnosis."[37]

Thus, Kosten and van Os wanted a clear verdict, which they considered to be within Bitter's particular line of expertise. He could, for instance, say: "In zones with a mean score of 5, the number of bothered people is so and so, and within a couple of years one would expect there to be so many people suffering from nerves, gastric haemorrhage, divorces etcetera that from a social-medical-psychological-physiological-ethical point of view one should speak of an untenable situation. Planning a residential district in such zones is irresponsible, regardless of putative economical or other advantages, while in current zones with a score of 5 the noise impact must be reduced or their use must be gradually changed."[38] Kosten and van Os put a lot of pressure on Bitter to design a system that would provide descriptions of what happened at which scale number. As they told him, "You can't escape to do it."[39]

Initially, Bitter tried to avoid complying with Kosten and van Os's demands. Bitter's boss, the institute's director, responded that the matter was actually an issue of policy.[40] A few months later, Bitter simply stated that the survey did not allow him to answer the questions Kosten and van Os had posed.[41] At last, however, he came up with the descriptions that were published in the final report. Bitter linked the permissibility of specific nuisance scores to the physical, mental, and social well-being of the population by stressing the main negative effects of sleep disturbance, communication disturbance, and fear. In his view, these factors could influence the population's well-being "considerably"—"disturbing environmental stimuli" could "harm" a person's well-being "to a greater or lesser extent." For instance, an annoyance score of 3, defined as a "predominantly moderate hindrance," which had B=45 put in the margin by handwriting, indicated the "boundary of permissibility," or the boundary of tolerance that "is reached for one third of the population." A score of 4, considered a "predominantly serious hindrance" (B=60, also in handwriting), however, was "not permissible," because "the boundary of tolerance is crossed for about half of the population."[42] Thus, only after prolonged pressure did Bitter provide the verdict that Kosten and his colleagues had asked for so many times. And although the B-values may have been added by Kosten or van Os as a comment on Bitter's document, or by Bitter himself in a discussion with Kosten and van Os, Bitter had suggested which score represented the boundary of permissibility. Kosten had finally found his Uncle Doctor: a man behind whom he could hide.

The authority of the "doctor" did not help to prevent Kosten's aircraft noise index from being adapted, however. What did survive from the committee report was Kosten's idea of surveyed zoning, the use of preferential runways, flight procedures and aircraft noise certification, and the installment of an Information Centre on Noise Nuisance at Schiphol Airport.[43] Yet Kosten's standard and the associated norms shifted in the course of time. Although the minister of transportation and water management claimed to accept the final report without reservation, he eventually set up "no less than five commissions for renewed study of the various aspects of aircraft noise" (Dierikx and Bouwens 1997: 193). One of these was asked to study the fairness of the 45 Ke norm. In 1972, it came up with a lowered maximum of 40 Ke for *new* residential areas, inspired, as it claimed, by (re)studying the Wilson report. At the same time, however, it suggested, unlike Kosten had done, the renovation of areas that already had a noise impact between 40 and 65 Ke, whereas it advocated that areas above

65 Ke should be declared uninhabitable.[44] In the 1970s, the officially proposed maximum for new residential housing dropped to 35 Ke, and a new map was made showing noise contours of 35, 40, and 65 Ke around Schiphol (Bouwens and Dierikx 1996: 306–307). This map and the related costs for soundproofing, renovating, and clearing housing affected the heated debate about the future of Schiphol. Groups actively fighting aircraft noise, such as the Foundation against Aircraft Noise Nuisance (*Stichting tegen geluidshinder door vliegtuigen*)[45] scrutinized the Ke-norms. Such groups were new social movements that, from the late sixties onward, sought to abate noise from a similar perspective from which environmentalists had sought to abate air, water, and soil pollution (Goodwin and Jasper 2003; Meijer Drees 1966). They came up with detailed criticism on the norms chosen as well as with alternative airport locations.

Still, the government started to insulate housing within the 65 zone—the zone of uninhabitability. The government did so with the help of Schiphol that claimed to assist those who did not want to leave (Bouwens and Dierikx 1996: 333, 382). Moreover, whereas Kosten had opted for counting aircraft noise above 70 PNdB (or 56 dB(A)), the policy that was eventually implemented did not count the noise of aircraft below 65 dB (Berkhout 2003: 17). In the early 1990s, Schiphol's maximum capacity came to be defined in terms of flight movements, passengers, and freight, provided that there would be no more than 10,000 residences in the 35Ke contour (Bouwens and Dierikx 1996: 389; Berkhout 2003: 15). The "final" delineation of this contour was settled in 1996—just four years before the European Commission decided to harmonize all European noise indexes (Berkhout 2003: 43).

In sum, the Dutch experts designed their own aircraft noise index after all, despite their time-consuming odyssey in search of an international solution, which was caused by disagreement among experts about which international organization should be in charge of setting standards. If two groups of countries fought each other within ISO about the basic unit of measurement to be used, dBA or PNdB, this struggle also suggested different cultures of control and legal contexts, including direct measurement versus computing, affordable surveillance over zoned areas versus preventing public complaints and preparing for civil lawsuits. This was one of the reasons why Kosten dismissed the NNI as an example to follow, since it used the "American" PNdB. The other reasons for rejecting it were the unit's lack of night penalties and exclusion of aircraft noise below a particular maximum. Problems with the German duration and night penalty system made the Dutch reject the German standard. Finally, the Dutch considered the local situation around Schiphol unique, as other countries had done for their own airports, and they decided to conduct a survey themselves, which led to their controversial definition of the proportion of the population that ought to be allowed to suffer from aircraft noise.

<h2 style="text-align:center">MECHANICAL AND PRAGMATIC OBJECTIVITY</h2>

The harsh language with which the Dutch rejected other countries' aircraft noise indexes, the coercive correspondence between Kosten and Bitter, Robinson's inconsistent stance on the issue of perfection, and the uncomfortable British refraining from revising the NNI become even more understandable, however, when we focus on mechanical objectivity and how it differs from and coexists with what I would call "pragmatic objectivity." The conflicts discussed in this chapter display a structural affinity in character that results, I claim, from the tensions between the ideal of mechanical objectivity and the logic of pragmatic objectivity.

In "The Image of Objectivity," Daston and Galison stress that our current notion of objectivity is "layered" and "mixes rather than integrates disparate components, which are historically and conceptually distinct" (Daston and Galison 1992: 82). One component of our notion of objectivity is the mechanical objectivity that favors mechanical registration of phenomena, the origins of which they situate in the late nineteenth and early twentieth century. In *Trust in Numbers*, introduced in chapter 6, Porter also uses the notion of mechanical objectivity for the late twentieth century and defines it as the "ability to follow the rules" that holds "professional as well as personal judgment ... in check" (Porter 1995: 4–5). Methods of processing numerical information with help of a computer are linked to this ability. The accompanying trust in numbers, as Porter claims, is dominant in fields with "insecure borders." Such fields include the applied sciences "that bear on matters of policy," because our democracies require objectified expertise rather than "seasoned judgment" in public affairs (Porter 1995: 230, 229, 7). Paradoxically, scientific communities that are most free from the pressures of political requests for advice, such as high energy physics, care the least about the explicit exchange of numbers. Yet now that the state increasingly supports science for "practical purposes," it "encourages the greatest extremes of standardization and objectivity, a preoccupation with explicit, public forms of knowledge" (Porter 1995: 229).

The story of applied acoustics and its search for a practical aircraft noise index would not have been out of place in Porter's important book. Yet, Porter's use of the notion of mechanical objectivity for most of late-twentieth-century science downplays Daston and Galison's insightful conception of objectivity as a layered notion. Since it is exactly the coexistence of distinct forms of objectivity that helps us to grasp why the disputes mentioned, all *within* the world of science, had such a confrontational character. Behind these clashes were the tensions between the ideal of mechanical and the logic of pragmatic objectivity. Mechanical objectivity, in my take of it, focuses on the registration of phenomena (in this case aircraft noise and responses to such noise) in whatever manifestation and *at whatever level* conceivable, that is, in its full scale of variability. Pragmatic objectivity, however, aims to create systems of numbers on the basis of which one can intervene (in this case, in the situation around airports) *at discrete levels.* So whereas Porter contrasts a trust in elite, professional judgment with a—now predominant—trust in publicly available and accountable numbers, I distinguish between trust in two forms of transparent quantification, yet with a difference in what the numbers are supposed to *do:* showing (continuity in) variety or creating discontinuity.

To elucidate my point I distinguish between regulative and coordinative standards. The latter are geared toward the compatibility of technological artifacts, the former toward the prevention of what are considered to be negative consequences of technology (Schmidt and Welre 1998: 120). When focusing on the aircraft noise indexes as regulative standards and analyzing their structure from this perspective, it becomes clear what these indexes, despite their differences, had in common. First, the indexes started out from the idea of *step by step* index numbers, such as 50-55-60 NNI, 72-77-82 Q, or 35-40-65 Ke, which were associated with similarly discernable levels of intervention in zones. Second, all the indexes incorporated a kind of zero-point of intervention, below which no intervention was deemed necessary. Third, and following the logic of the first two characteristics, two groups of citizens, those always disturbed and those never disturbed, were declared irrelevant to the systems of intervention. The use of pragmatic objectivity thus dismissed "extreme" responses to phenomena in order to enable discrete levels of intervention. The perception of sound of those who fell within the two extreme groups of respondents did, so to speak, not make sense in light of pragmatic objectivity.

The use of aircraft noise indexes implicated both forms of objectivity, in a superimposed manner, but noise experts struggled with their coexistence. Robinson strove for supranational intervention at the expense of perfection, thus stressing pragmatic objectivity, but kept defending jury testing as the best way to register the array of responses to aircraft noise, a process that required the use mechanical objectivity. When the British Noise Advisory Council decided to hide its doubts about the substantiation of the NNI to the British public, it was afraid that it would lose pragmatic objectivity if they stuck with mechanical objectivity. Their response even entailed a partial *blackboxing* of the aircraft noise index, and thus a loss of accountability. The same trade off occurred in the case of the Netherlands. Kosten, the head of the Dutch effort to establish a standard unit, tried to distance himself from drawing the lines between acceptable, acceptable-under-conditions, and nonacceptable effects of aircraft noise, but he pressed Bitter to do this for him. He thus forced Bitter into the double role of being the observer of the infinite variety of aircraft noise responses *and* the one who to had to discern discrete intervals in noise levels.

CONCLUSIONS

Scientists have, from the late 1950s onward, persistently promoted international solutions to the aircraft noise problem. That it took so long to find a common standard was surprising in so far as many of the conditions that could have enabled the smooth circulation of public problem definitions and policies, as listed by public problems' theorist Joel Best and others, were already in place. These were the close international contacts between the acoustical experts involved, who frequently met each other at conferences or through their involvement in international bodies (helped by the aircraft they discussed!), the theorization of the aircraft noise problem, and the pervasive trend of increasing international air traffic in the West. Best has also mentioned the factors that may hamper diffusion of problem definitions and solutions, including xenophobia and all kinds of institutional differences between countries, such as dissimilarities in the extent of governmental centralization, the character of law, or the organization of medicine (Best 2001).

If among the various countries involved there is no evidence of xenophobia as a factor in the aircraft noise issue, institutional differences definitely contributed to each country going its own way. Examples include specific research traditions, such as the jury of listeners' approach used by the British National Physical Laboratory and the tradition of using laboratory studies into the effects of industrial noise that the experts advising the German government employed. Also relevant were the specific legal contexts involved, such as the possibility of citizen's claims on the basis of common law or the putative need for surveillance. Such contexts, in turn, required particular policies that eventually even played a role in the arena of international organizations such as the ISO, just one of several agencies and bodies that tried to set aircraft noise standards. These bodies, however, proved hardly willing to grant one of the other parties the right to regulate. And after countries had adopted indexes of their own, their embedding in national noise control regulations and maps hampered the introduction of an international index. The United Kingdom even held on to its index for fear of the public's distrust of science-based standards in general. Moreover, both the British and the Dutch choices were marked by underlying tensions between mechanical and pragmatic objectivity.

These conclusions imply that introducing European Union-wide aircraft noise indexes at the expense of the existing ones, as the European Union initiated in 2000, will not prove to be an easy trick. Choosing a particular standard is one thing; embedding it in the varying research and policy traditions of the member countries is quite another. This can be inferred not only from the obduracies highlighted by my argument, but also from today's continuous disagreements on how to gather the information that should feed aircraft noise indexes (Ten Hoove 2005). This as well as previous work on the history of standardization suggest that there is a world to gain after standards have been considered.

Two last remarks are called for. As we have seen, the pragmatic objectivity behind the aircraft noise indexes dismissed the experiences of those always and never disturbed by aircraft noise, reduced the transparency of the indexes (because of its struggle with mechanical objectivity), and led, once again, to a spatial framing of the problem. Aircraft transcended physical borders very easily, but the abatement of its noise became territorialized by drawing sharp boundaries between zones confronted with different levels of noise impact and with different forms of intervention. What this meant for the experience of control by citizens will be discussed in the chapter 8. This chapter, the last chapter of this book, will also reflect on the role of economics in public problems of noise. Aircraft noise was due to a booming business, and thus of economic growth. Yet, we should not forget that fighting acoustically induced fatigue in the aircraft industry, enhancing fuel efficiency, and granting subsidies for new types of airplanes have stimulated the silencing of aircraft more than the array of national aircraft noise indexes produced in the 1960s. We should therefore rethink the putative opposition between economic progress and environmental concerns.

A SOUND HISTORY OF TECHNOLOGICAL CULTURE

Nooit meer stil
Men zal nog krijgen dat het nooit meer stil is
en dat voortdurend 's nachts en altijd de verdoemde rotmotoren
om je kop ronken
zodat het zoemen van machines voortaan altijd
door de lucht gaat
knalpottende raketten rondstotteren
en dat geen plekje hei of wei
geen kilometer aardig lenteland meer vrij is
van hun gezoem waarmee ze alles verpesten
Het vinden en verbeteren zélf is mooi
maar het gevolg is toch maar een
massa doelloos tuig
vlieg-tuig

[*Never ever silent*
Once we'll witness that it is never ever silent
and that, at night, the damned rotten engines
will always roar around your head

henceforth, the humming of machines
is in the air for ever
stuttering rockets stammer around
and no spot on heath and meadow
no acre of pleasant spring land will be free
from the buzz they spoil everything with
Inventing and improving is good in itself
yet the consequence still is a
mass of senseless craft
air-craft]
— *(Hanlo 1958/1946: 34)*.

A TRAGIC HISTORY

In his study on the culture of public problems, Joseph Gusfield aimed to show that the American definition of the drinking-driving problem could have been otherwise. Such an endeavor left the sociologist, he claimed, with two alternative positions. One was the *utopian* stance, in which "the uncovering of the ephemeral character of the dominant perspective provides the occasion for the creation of a newer and better one." The other, the one preferred by Gusfield, was the *olympian* perspective, which is "skeptical of all perspectives" (Gusfield 1981: 193).

Gusfield did not mention a third position, the one that has been the approach in this book: that of *tragedy*. By showing how the public problems of noise changed over time and only temporarily acquired sufficient drama, robustness, and connections to be tackled thoroughly, the history of noise problems has largely been presented as a tragic story. Despite the many attempts to control noise, noise still features prominently on the Western world's public agenda. As Hayden White (1990) has pointed out, historians employing the plot of the tragedy write in a *romantic* style by consequently stressing the role of fate in what has happened. They tend to underline the significance of structural forces pushing developments outside the reach of individuals. This book has, indeed, often followed this storyline in a slightly different vein: by showing how early interventions in the public problem of noise constrained the options subsequently available. This process of framing new noise problems in terms of older strategies of noise abatement, as bricks set on each other, has had two significant consequences: the rise of a paradox of control, and an intriguing spatial focus in noise abatement. Both consequences contributed to the persistence of the public problem of noise.

Before discussing these issues, this chapter will reappropriate, in the light of the previous chapters, the three most common answers to the question of why noise abatement has been such a protracted affair. These answers include the assumptions that noise is a product of technological progress and that we have valued innovation over silence; that the subjectivity of hearing dashes all hope of controlling noise; and that sound can never become of true value and significance in a visually oriented culture. Subsequently, I will analyze the establishment of the British, German, and Dutch noise abatement laws in the 1970s in terms of the structure of the solutions chosen. This will enable me to show in what ways these solutions drew on older notions of sound, legislative traditions, and instruments of measurement, all of which led to a spatial focus in noise abatement and a paradox of control. Finally, this chapter will point out some lessons that have been learned about how to write a sound history of technological culture and propose a few alternative approaches to noise abatement that take both its past tragedy and occasional successes into account, that is, ecological modernization, sensibility in legislation, and complaining in style.

THREE ARGUMENTS REVISITED

What can be proposed against the argument that noise problems have remained on the public agenda because we, as a culture, have accepted noise as the necessary companion to progress? It sounds like such a convincing claim, to which Jan Hanlo's poem above gives a new voice. The claim's line of reasoning, however, leaves out three important issues that have been discussed in previous chapters. The first is that noise has indeed often been associated with progress, not merely as a *by-product* of economic growth, but as a *sign* of strength, power, and progress. That the physicians who tried to persuade factory workers to use earplugs did not immediately succeed in their endeavor was due largely to the fact that they did not bear in mind the positive meanings loud sounds could have to the workers involved. Those taking up noise abatement could have benefited from taking such positive connotations seriously. This implies that the obduracy of noise problems not only follows from taking the thick and thin together, as George Trobridge nicely phrased it in 1900, but also from not fully acknowledging the deep-rooted symbolism of noise.

A second issue that weakens the claim that noise problems have been insufficiently tackled because noise is the self-evident escort of prosperity is that particular technologies have become less noisy over time precisely because engineers considered noise as evidence of the inefficiently working machines. In these cases, noise was seen as something that might threaten progress, as the histories of industrial and aircraft noise have shown. The biggest steps in reducing the sound level of aircraft came from fighting acoustically induced fatigue, enhancing fuel efficiency, and fostering the design of new types of aircraft, that is, from boosting technological, military, and economic progress. Nor should we overlook that industrialization and economic change brought an end to many highly noisy activities, such as the businesses of blacksmiths, millers, boilermakers, and vendors. The same holds for many noisy artifacts: rubber tires, new car constructions, and asphalt replaced iron-girded wheels, clattering steel frames and cobblestones, reducing friction and thus noise.

A third issue relevant for assessing the economic-progress-at-the-expense-of-environmental-concerns argument involves rethinking the assumption that noise has been so easily accepted. In fact, many new machines encountered protests against their sound. Such sounds stood out in people's perception exactly because of their novelty: their innovation expressed what the general public had not expected to happen. To many citizens, new sounds fell from the sky—literally, in the case of aircraft—which contributed to their perception of technology's sound as noise. This realization raises a critique on Raymond Murray Schafer's dichotomy between a hi-fi and a lo-fi society. Industrialization has indeed changed the keynote of our society in the sense that a blanket of continuous sounds is masking other sounds. It reduces, in Schafer's terms, the high fidelity, or the signal-to-noise ratio, of our sonic environment. Yet we still perceive high fidelity as new sounds catch our attention. The previous chapters have shown that new sounds usually triggered fierce discussions about their quality. Their acceptance took time and effort.

We cannot ignore the fact that the Western world inhabits much more sound-producing equipment today than in the late nineteenth century. Appreciating economic progress, however, does not necessarily imply an acceptance of noise. It was once widely believed that the commercial interests of the cigarette industry would always win against those interested in saving the health of smokers. But now that we have redefined health problems in terms of economic costs and benefits, and have identified smoking as a serious danger to health and, thus, prosperity, smoking has been banned from a growing number of public areas. That noise has not been defined as a serious danger to public health is the issue that needs to be explained, in a manner different from merely referring to our obsession with prosperity.

What, then, about the argument that our acknowledgment of the subjectivity of hearing has hampered us in our search for a long-term solution to the noise problem? To be sure, we cannot live in a world without sound: it is integral to our lives in many crucial ways. Yet, the observation that some people hate the sounds that others love has not continuously undermined the case for noise abatement. Ever since noise has been featured on the public agenda, people have claimed, often with deep regret, that the response to particular sounds is highly varied. But in the late nineteenth and twentieth centuries, sensitivity to noise was, at least in the case of the intellectual elite, seen as evidence for having a refined, delicate, and cultivated mind. In the elite's view, those who felt undisturbed by the chaos of city sounds exhibited their membership in the lower, vulgar classes. A civilized and mentally healthy population was therefore in need of tranquility.

These were the convictions that energized the first wave of antinoise campaigns. They even explain the partial success of the early campaigns, since the association of societal order with silence enabled a discourse coalition with those who focused on reorganizing the traffic flow. Although the coupling of tranquility with intellectual and social refinement blocked the road to other coalitions, with labor unions for instance, the idea that sound perception was subjective did not hamper noise abatement itself. On the contrary, presenting oneself as sensitive *and* socially or intellectually elite initially empowered the cause of noise abatement. That the tone and leadership of the first wave of noise abatement societies had a right-wing flavor contributed to this process of authorizing the antinoise movement. This situation differed vastly from the more leftist and environmentalist atmosphere of the second wave of noise abatement activities in the 1960s and 1970s.

It was only in the course of the 1930s, when experts in psychoacoustics started to medicalize and pathologize one's sensivity to noise, that the notion of the subjectivity of hearing began to hamper particular forms of noise abatement. Now that a person's nonadaptation to noise was increasingly seen as a symptom of being incapable of concentrating on the task at hand—a shortcoming presumably caused by a person's deeper problems—being sensitive to noise was not something to be proud of anymore. Consequently, noise could not be banned by implicitly referring to the high status of those disturbed by it. At the same time, the right to have the equipment with which one could control sound and produce music came to be democratized from the right of a few to the right of the masses. Such music might disturb neighbors, but after the Second World War, scientists and authorities granted citizens the right to express themselves, at least once in a while, through making noise.

This shift in the balance of power between the classes and the appreciation of the subjectivity of hearing paralyzed some types of noise abatement. Public education that simply broadcast the message "get civilized, keep silent" became increasingly embarrassing and thus less convincing and had to be replaced by forms of public education that focused on people's skills in confronting those who caused the sounds they disliked. This change in focus indicates that the subjectivity of hearing has had different values attached to it at different times, and that it has only reduced the chance to strike noise off the list of Western public problems since the late 1930s. In the nineteenth century, the subjectivity of smell had, alongside the putative danger of stench, lent force to the far-reaching interventions the French elite took in abating noise, which extended even into the (private) home itself. The wide public discussion of noise entered the stage much later than the heated public debate about stench, however. At the end of the 1930s, direct intervention into the home and on the domestic behavior of citizens, whether with the help of portable noise meters or not, was not considered legitimate anymore, because of shifting class relations and a new appreciation of sound perception. And since psychoacousticians increasingly considered the effects of noise on the mind and bodies of individuals to be temporary problems, to which all but the mentally overcharged could adapt, they downplayed the health problems formerly associated with noise. The notable exception was the medical acknowledgment of noise-induced hearing loss, but the effects on people's blood pressure and concentration were seen as more transient problems.

One argument still remains to be revisited: the claim that we should blame the sensory priorities of our culture for the neglect of concerns about noise. The idea behind this claim is that the West has overlooked the problem of noise because of its cultural preference for the eye at the expense of the ear. My position is, in fact, the opposite. Our culture has gradually become increasingly dependent on the eye as a *consequence* of the occasional successes in noise abatement. As chapter 4 has shown, several solutions to the problems of traffic noise involved making things visible that had formerly been audible and preventing auditory signaling by creating surveyable cities. The broadening and straightening of roads, the increasing use of stoplights, and teaching the public to flash their headlights instead of honking their horns when approaching an intersection are all illustrations of this trend. Other examples include the use of nameplates to identify streets, bans on honking at night, and the replacement of bells for announcing the arrival of ships, trains, and the hours of the work day by (silent) schedules and notice boards.

As chapter 1 argued, finding encouragement and confirmation in Jonathan Sterne's work, adherents of the visual culture argument consider the ear the morally better sense. They suggest that the emancipation of the ear is the sine qua non for any serious undertaking of noise abatement, whereas I would stress that noise abatement has been strongly assisted by our ability to see. Such an argument neither implies nor necessitates a culturally privileged position for the eye in modern Western culture, however. As the chapters 5 and 6 have illustrated in abundant detail, producing, controlling, and listening to music have become highly valued and increasingly prevalent activities, a fact that underscores a rise rather than a decline in the cultural significance of the ear in the twentieth century. Music's new cultural prevalence has certainly created new problems, but they cannot, I argue, be explained by our culture's putative sensory priorities. To explicate the history of public problems of noise, we need an alternative narrative: a sound history of technological culture.

THE STRUCTURE OF SOLUTIONS

In 1931, the British physicist G. W. C. Kaye beautifully expressed one of the later ideas of the anthropologist Mary Douglas by referring to noise as "sound out of place." A sound could come to be out of place, he claimed, by its "excessive loudness, its composition, its persistency or frequency of occurrence (or alternatively, its intermittency), its unexpectedness, untimeliness or unfamiliarity, its redundancy, inappropriateness, or unreasonableness, its suggestion of intimidation, arrogance, malice, or thoughtlessness" (Kaye 1931b: 443, 445).

Today, we know that not all of these aspects of sound have been the focus of intervention in western laws concerning noise abatement. The questions that arise, then, are which ones *did* become the subject of intervention, and why they and not others? To answer these questions, I will analyze the laws on noise abatement established in the United Kingdom, the Netherlands, and Germany in the 1970s in terms of their structure of solutions, and will explain this structure on the basis of the previous chapters.

This book has shown that the large majority of the interventions focused on the individualizing, objectifying, or materializing the problem of noise. Every type of intervention entailed a distinct approach that focused on the source, transmittance, or reception of noise. *Individualizing* the problem of noise, for instance, occurs in solutions that center on the individual choices of citizens and focus on changing the behavior of individuals through education. Examples of individualizing interventionary schemes are the distribution of leaflets asking drivers to use their horns sparingly (source), requests to close windows when playing the radio (transmittance), or campaigns to persuade people to use earplugs (reception). *Objectifying* solutions to the problem of noise include all interventions that set time slots to or limits on the level of sound emission and immission, and always imply measurement—of time, of sound, or both. Objectifying interventions encompass establishing the maximum levels of sound emission of cars and aircraft (source), setting obligatory standards for the quality of sound absorption and the insulation of walls (transmittance), and establishing norms that define the maximum levels of noise allowed to have impact on the residents who live near highways and airports (reception). And, finally, *materializing* strategies to the problem of noise include all technical constructions that reduce the emission, transmission, and immission of noise, such as silencing automobile exhaust systems and engines (source), insulating walls and floors (transmission), and creating earplugs or noise barriers (reception). Even playing music on the shop floor is a material way of reducing the reception of noise.

It is important to note that individualizing interventions (use earplugs!) can have material aspects (use *earplugs*), while their main strategy is to focus on changing a particular behavior (*use* earplugs). Another point that is relevant for the argument that follows is that individualizing, objectifiying, and materializing the problem of noise may involve using different types and scales of collective action: from campaigns undertaken by private initiatives to regulations established by local or state authorities. And to recall Gusfield's work, different interventions follow from particular definitions of a public problem—in terms of its causes and solutions—by the (often self-proclaimed) owners of the problem, and they always express particular conceptions of who is responsible for the solution. In the case of individualizing interventions, the responsibility for solving the problem is attributed to the individual citizen. In materializing solutions, responsibility is often transferred to those who make the technology or instrument involved, usually engineers. In the case of objectifying interventions, it is the municipial, state, or federal authorities who are held responsible.

Which of these interventions were mentioned in the noise abatement legislation in the 1970s? How did these relate to the ways in which sound could be "out of place" as Kaye had argued fifty years earlier? Which forms of intervention, if any, were entirely new? And how can we understand the sum of interventions from the history of the public problems of noise discussed in the previous chapters? To answer these questions, I will start by comparing the British Control of Pollution Act of 1974, the German Federal Immission Protection Act (*Bundes-Immissionsschutzgesetz*) from the same year, and the Dutch 1979 Noise Abatement Act (*Wet Geluidhinder*) in terms of their main intervention strategies. All these laws were tokens of a legislative framework that set, at different levels of detail, the scope of future measures to be taken by local, regional, and national authorities.

The British Control of Pollution Act repealed both the Noise Abatement Act of 1960 and all local by-laws that contradicted its content (Kerse 1975: 157). The new act covered not only the problems of noise and vibration, but also those of air and water pollution. With respect to noise, it empowered local authorities to "declare a 'noise abatement zone' for an area within their district," and thus "to take action to prevent noise ... of an intensity higher than the level prevailing at the time the zone was declared." In addition, local authorities were entitled to prescribe how work on construction sites could be carried out in a tranquil manner and to "require [current] noise levels to be reduced" (Garner 1975: 2). Unlike the Noise Abatement Act, the Control of Pollution Act applied even to statutory undertakers such as British Railways, with the significant exception, as we have seen, of the airlines. What remained from the Noise Abatement Act was a ban on the use of loudspeakers in streets at night (between 9 p.m. and 8 a.m.), and the fact that local authorities were not the only actors able to set the noise-abatement ball rolling. A citizen could still complain at a magistrate's court "on the ground that in his capacity as occupier of the premises" he was "aggrieved by noise amounting to a nuisance." A court might act on such a complaint if it was "satisfied that the alleged nuisance" existed (Garner 1975: 97).

The act's new zoning procedure necessitated that local authorities measure the level of noise emanating from any premises within the zone and record them in a "noise level register." Subsequently, anyone who exceeded the noise level was considered punishable, unless they had been given prior consent to do so by the local authority. Local authorities were required to inspect the zones for noise nuisance regularly. In the case of an offense, they had to issue a notice requesting the abatement of the nuisance or a notice informing the specific steps to be taken in order to achieve its abatement. Had the noise been produced "in the course of a trade or business," it was defensible if one could prove that "the best practicable means" had been used for "preventing, or for counteracting the effect of, the noise" (Garner 1975: 95). The "means" in the legislation referred to the design, construction, installation, operation, and maintenance of buildings, acoustic structures, plant, and machinery, while "practicable" meant that one could account for local circumstances, the current state of technical knowledge, and any financial implications (Garner 1975: 111). In order to disseminate such technical knowledge, the government could publish codes of practice for "the purpose of giving guidance on appropriate methods ... for minimising noise" (Garner 1975: 110).

At first sight, the requirement that the "best practical means" should be used was similar to "technological state of the art" (*der Stand der Technik*) demanded by the German Federal Immission Protection Act. Just as in the British case, in Germany the legislation was aimed at abating several kinds of environmental problems including noise, vibration, and air pollution. And as the British had done, the Germans did not specify their immission and emission limits within the law, but allowed the announcement of future specifications. With respect to noise specifically, the law regulated the condition and use of vehicles, trains, aircraft, and ships, and the (re)construction of streets and railways. Yet the heart of the matter was that in order to acquire a license, businesses had to prove that they could carry on their work without causing environmental damage and dangers, or even considerable nuisances to the neighborhood and wider community, and had to take steps to reduce noise that would live up to the technological state of the art. Federal authorities also had the right to define these technical requirements in detail in the future, and to set particular noise emission limits. The federal government took care of the abatement of noise from industries, vehicles, and roads, whereas the German states (*Länder*) had the responsibility of controlling the noise made by humans and animals.

Despite the many similarities between the British and German legislation, the "technological state of the art" in the German law meant something different from the "best practical means." The steps taken to reduce noise nuisance by German businesses had to meet the quality of the best solutions—technological or organizational—of the world. The measures thus had to exceed local customary practice and to focus on the frontier of technological development. The only exceptions allowed were situations in which a particular measure would create a "gross imbalance" between the financial investment needed and the noise reduction that was attainable (Jarass 1983: 50).

The German Federal Immission Protection Act repealed a substantial part of the trade code (*Gewerbeordnung*) that has been discussed in chapter 3, including a section to which the Technical Guide for the Protection against Noise (*Technische Anleitung zum Schutz gegen Lärm*) belonged. The Technical Guide was to remain valid, however, insofar as the new law did not contain similar detailed prescriptions for noise control (Jarass 1983: 381–382, 467–468). This meant that, for the time being, the immission norms established in the Technical Guide would remain binding. Since 1968, these had been 70 dB(A) for industrial areas, 65 dB(A) for areas that were primarily industrial, with a night limit of 50 dB(A), and 60/45 dB(A) for areas that were neither strictly industrial nor strictly residential—with the night defined as the hours between 10 p.m. and 6 a.m. For purely residential areas, the maximum allowed immission of noise had been 50/35 dB(A), and for spas, hospitals, and nursing homes 45/35 dB(A).

Both the German and the British new legal framework for noise abatement repealed only part of the earlier legislated interventions. The British, for instance, kept their Road Traffic Act and Motor Vehicle Regulations from 1972, both of which were revisions of the ones discussed in chapter 4, as well as the Building Regulations, also from 1972, the Noise Insulation Regulations (1973) for the soundproofing of residential housing near new roads, and a Land Compensation Act (1973). In contrast, the Germans had adopted the Act for the Protection against Construction Noise (*Gesetz zum Schutz gegen Baulärm*) in 1965, and in 1971 replaced it with the Act for the Protection against Aircraft Noise (*Gesetz zum Schutz gegen Fluglärm*), which provided citizens with compensation of damage caused by aircraft noise (Klosterkötter 1974: 9).

The Dutch Noise Abatement Act excluded aircraft noise, and left the protection of employees from noise, the regulation of vehicle and road noise, the ban of noise near churches on Sundays, and the noise of neighbors to other jurisdictions of the law or to forms of education. The 1979 act, however, did offer the option of regulating the maximum sound emission from an industry. But the core of the Dutch Noise Abatement Act, like that of its British and German predecessors, was the link between spatial planning and noise control, in this case, the confinement of particular levels of noise immission to particular zones. Unlike the British and the German legislation, however, the Dutch Noise Abatement Act specified the details of these levels in the law itself.

The Dutch Noise Abatement Act distinguished three types of zones: zones around industrial areas, zones along highways, and zones along railways, tramways, and subways. With the establishment of these zones by regional and local authorities, planning within them would become impossible without prior acoustic inquiries and authorizations: one simply had to take the maximum allowed immission levels into account. These limits came to be expressed in terms of Leq, the equivalent sound level defined as the average sound levels at a particular place in a particular period of time as measured in dB(A). The Noise Abatement Act defined the width of the highway and railway zones. The industrial zones, however, were the jurisdiction of municipal councils.

Outside these industrial zones, the maximum noise impact allowed at the facades of residential housing was 50dB(A). Inside the home the limit was 35 dB(A), provided that the windows were closed. New housing in industrial zones had to meet the same criteria. Existing housing, however, could do with 55 dB(A) or even 60 dB(A), for instance, in cases where industrial areas were also bothered by traffic noise (Neuerburg and Verfaille 1995: 370–375). One novelty in the legislation was that the provincial authorities were allowed to allot "silence areas," in which the level of man-made sounds did not disturb (or only scarcely disturbed) the perception of natural sounds. One traditional aspect of the law, however, was the option to create additional immission limits for "noise sensitive buildings," such as schools, libraries, hospitals, health resorts, nursing homes, and institutions for the mentally retarded, as well as offices, museums, theaters, and concert halls (Neuerburg and Verfaille 1995: 360).

This quick scan of the noise-abatement legislation adopted in the 1970s shows that all new interventions were variations on those identified in earlier episodes, yet they were increasingly focused on objectifying and materializing the problem of noise. It was notably the combination of zoning, setting sound level limits, and imposing time slots through night norms—a trio that we have already seen in the discussion of aircraft noise—that became the crux of the noise abatement legislation during this period. Moreover, standardized zones and maximum allowed levels of sound pushed the role of individual citizens' complaints aside, although individual complainants continued to have a part to play in the British legislation. It is also remarkable that, with the exception of the Dutch Noise Abatement Law, the criteria for the quality of noise control had explicitly been delegated to technological development. The traditional islands of silence in the West, however, remained firmly in place, although places of entertainment were now also considered worthy of special protection. In addition, on the continent, controlling mechanical sound became largely the jurisdiction of the state, whereas lower authorities took care of nonmechanical sound.

If we subsequently compare the structure of solutions with the causes of noise mentioned by Kaye, we are able to see which causes have been included and excluded from the noise legislation during the 1970s. Zoning, setting sound level limits, imposing time slots, and silencing equipment were definitely in, and are linked to Kaye's "excessive loudness," "frequency of occurrence," "untimeliness," and "redundancy" of noise. The public education of citizens that, as we have seen in chapter 6 and earlier chapters, became the responsibility of the noise abatement societies has taken hold of the "inappropriateness" and "unreasonableness" of noise, as well as of the "intimidation," "arrogance," and

"thoughtlessness" presumed to be behind making noise. Strikingly, "intermit-
tancy," "unexpectedness," and "unfamiliarity" are not part of the solutions at all.
How can we understand these trends, and what are their consequences?

THE PARADOX OF CONTROL AND THE SPATIAL FOCUS OF NOISE ABATEMENT

Let me now examine the historical trajectory of the changing definitions of
noise problems, outlined in table 8.1, while also keeping in mind the structure
of solutions discussed in the previous section. For every problem of noise discussed in this book—industrial noise, traffic noise, neighborly noise, and aircraft
noise—the table indicates the problem owners during each particular episode,
as well as the problems' causes, consequences and victims, solutions, and responsibility defined by the problem owners. As to the problems' solutions, I have
distinguished between public education (an individualizing strategy), "limiting,"
which refers to imposing limits on the maximum levels of sound emission and
immission allowed (an objectifying strategy), zoning and establishing time slots
(objectifying strategies), and silencing (a materializing strategy). The vertical line
in the table indicates shifts over time. In addition, the right-hand column represents the theoretical notions that helped me explain the outcome of the debates
for each of the various public problems discussed in this book. (Table 8.1.)

We can now identify several trends in the evolution of solutions to public problems of noise. While materialization through technical forms of noise
reduction has been a recurrent strategy of intervention, individualization and
objectification have changed in significance and content over time. As to the
first shift, particular noise problems have been delegated to public education, and
have thus been individualized, which was the case with the noise of neighbors.
Other public problems of noise, however, such as the issue of industrial hearing loss or traffic noise, have been deindividualized and gradually merged with
another major shift: the objectification of problems through zoning, imposing
sound level limits, and establishing time slots through technical forms of noise
reduction. And whereas zoning and "timing" are long-standing strategies of

Table 8.1 Overview of changing definitions of noise problems

Object of discussion and regulation	Problem owners	Problem causes	Problem consequences and victims	Problem solutions	Problem responsibility	Theoretical issue
Industrial noise	Politicians (Conservative)	Industrial businesses (Liberal)	Loss of enjoyment of land property	Withholding permissions — Zoning	Politicians — — Local authorities	Conflicts between problem dramatizations and definitions
	Lay people	Intrusive sound	Loss of income and wellbeing citizens	Withholding permissions	Local authorities	
	Engineers	Friction	Loss of energy in machines	Reduction of friction=> silencing	Engineers	
	Physicians	Lack of knowledge	The deafness of workers	Public education (earplugs) — Limiting	Workers — — Politicians	

						Discourse coalitions
Traffic noise	Intellectuals and activists	Chaos of sound and barbarism	Decline in health of elite	Public education and silencing	Noise abatement orgs and car-driver orgs and engineers	Dequantification and depoliticization
	Scientists	Level of sound	Decline of health of the masses	Limiting and silencing	Politicians and engineers	
Neighborly noise	Local politicians (nonleft)	The character of mechanical instruments	Being forced to listen in	Local ordinances	Local authorities	
				Public education and building regulations	Noise abatement orgs and politicians	
Aircraft noise	Scientists and politicians	The booming business of aircraft	Decline of health of residents near airports	Zoning, limiting, timeslots and silencing	International bodies and engineers	Pragmatic objectivity

intervention, as the islands-of-silence tradition shows, setting sound level limits entered the history of noise abatement only in the 1920s, after the introduction of the decibel. In the aircraft noise indexes and the interventions following their establishment, zoning, timing, and setting limits came to be combined—a combination that had increasing significance, as my analysis of the noise abatement legislation of the 1970s has demonstrated.

Let me first discuss the context and consequences of individualization. In chapter 6 I argued that the changing position of the subjectivity of sound perception, the democratization of playing music, and the rise of the right to make noise made neighborly noise into an issue that was largely beyond law. By this I mean that scientists and politicians came to see the noise of neighbors as a problem that could be confined through building regulations, but not by imposing nation-wide limits for the maximum levels of neighborly sound, or by stimulating the use of noise meters in settling disputes between neighbors. In contrast, citizens had to be educated in how to control their own noisy behavior, or even in accepting incidental noise, and in how to approach their noisy neighbors in a prudent manner. As I have claimed, following Nikolas Rose, this strategy of individualization has been highly functional in liberal democracies. Such democracies ask for the rational, expertise-based exercise of power *and* for autonomy in private space. Having professionals make adults into skilled mediators for solving neighborly disputes about noise responds perfectly to these needs, and implies a partial depoliticization and dequantification of the problem of neighborly noise.

This type of analysis interestingly matches the "informalization-thesis" that has been promoted by the Dutch sociologists Herman Vuijsje and Cas Wouters (1999). Their thesis, which was inspired by the work of Norbert Elias, begins by arguing that the process of civilizing the masses in Western culture has followed a path from punishment to internalized norms, that is, from the use of external force (*Fremdzwang*) to a focus on self-control (*Selbstzwang*). In the course of the 1960s, however, all fixed codes for behavior, external or internal, came to be contested. Instead of depending on unambiguous guidelines for behavior, citizens increasingly aimed at self-realization and a highly personalized lifestyle. This change was due largely to a decrease of social and psychological distance in our daily lives and an increase in the interdependency of persons from different classes.

Yet, as Vuijsje and Wouters claim, this new ideal of self-realization led not to large-scale uninhibited behavior, but to an "informalization" of social situations. This informalization in social relations is manifest in the rise of informal, fine-tuned, and complex codes of behavior, tailored to the situation at hand. Inviting one's boss over to dinner, for instance, can mean that one has to create a formal atmosphere, but by no means necessarily, since the final choice depends on one's particular relation with one's boss. Such situations ask for a mix of highly flexible *and* controlled types of behavior, notably since expressions of unveiled disrespect for other people are taboo. Social encounters therefore continuously require choices: Am I on a first-name basis with this person? What should I wear in this particular situation?

In neighborly disputes about noise, a similar flexible attitude came to be required. One cannot fight with one's neighbor, but instead one should display a refined set of skills for negotiating with them. Nowadays, Maastricht police officers recommend to residents who are bothered by their neighbors' noise to try to talk to their neighbors first before asking for mediation by a police officer (Politiewijzer 2005: 36). Yet, as mentioned in chapter 6, residents express a growing fear of approaching their noise-making neighbors, probably because, as Vuijsje and Wouters made clear, showing one's lack of respect for another's way of life is simply not done. This means that citizens are increasingly left to their own devices. And now that we "domesticate" an increasing number of spaces, including our gardens, by decorating and furnishing these in a manner similar to our living rooms, even with outdoor audio equipment (Göttgens 2005: C2), we may expect the boundaries of auditory privacy to become contested even more often than in the past.

Yet, while some noise problems have been individualized, others have been objectified through zoning, by imposing sound level limits and creating time slots. Chapters 4 and 7 clarified how this happened. Only gradually did the introduction of the decibel and noise meters change the definition of noise from a problem expressed in terms of the *chaos* of different sounds to a problem expressed in terms of the *level* of sound. In fact, this notion could only be fully acknowledged as practical after noise abatement policies had reduced the variety of sounds in the street, and had thus made the measurement and regulation of noise in terms of decibels easier. Moreover, with the emergence of the mass consumption of cars, traffic noise became less commonly attributed to the behavior of particular groups of people—and thus with public education—and more often conceptualized as a collective phenomenon. Subsequently, the old strategy of restraining noise to or banning it from particular zones in space and hours in time came to be linked to setting particular limits to maximum allowed levels of sound emission and immission. Moreover, chapter 7 showed how scientists had to opt for the logic of pragmatic objectivity—which lead to tensions with their ideal of mechanical objectivity—when attempting to advise governments about the maximum allowed impact of aircraft noise. In doing so, politics and science together quantified the problem of noise, following a general trust in numbers and explicit forms of knowledge, but, at the very same time, blackboxed the indexes and zones to which the discrete steps of intervention came to be linked.

In the case of noise, both liberal democracies' trust in numbers *and* their respect for the autonomy of private space and citizens' self-realization have affected the choice of interventions, leading to the individualization of neighborly noise and the objectification of the other noise problems discussed in this book. It is important to note, however, that we tend to redo our history of noise as soon as new technologies enter the scene. With the introduction of the cell phone (or mobile phone) in the 1990s, people started using their phones while traveling by train in the 1990s. The first response by those forced to become "unwilling eavesdroppers" was that this noise unjustly transgressed the boundaries between the social classes (Agar 2003: 71). They considered cell phone sounds as socially disrupting sounds, as sounds out of place, as had happened in many earlier cases, when old sounds were new. It is therefore not surprising that the first publicly proposed solutions focused on calling for civilized cell phone behavior, or "*mobieliquette*" (Mulder 1997). Yet, in the twenty-first century, now that almost all of us possess a cell phone, we have delegated the problem to zoning once more: trains now have nonphoning zones where they had nonsmoking zones in the twentieth century.

Today, we tend to change from individualizing to objectifying the problem of noise in cases where we consider the space in which sounds bother us as public, and the source of sound as collective. In contrast, we prefer individualizing over objectifying when we consider the space in which sounds disturb us as private and when we attribute the source of the sound to particular groups of people. Subsequently, we phrase new noise problems in terms of older individualizing and objectifying strategies, thus stacking interventions as bricks on top of each other. Given such processes, I expect that the domestication and privatization of the garden, the extension of the living room and its audio-sets to space outside, will lead to new neighborly disputes about noise. And the restoration of tranquility in the garden will probably end up being the personal responsibility of citizens.

The processes that bring some noise problems under the objectifying regime and others under the individualizing regime, and the way in which new noise problems are embedded in strategies used to tackle older ones, have had two significant consequences that contribute to the persistence of noise on the public agenda: the rise of a paradox of control and the establishment of a spatial focus in noise abatement. I will first unpack the notion of a *paradox of control*. Scientists and politicians have increasingly promised to reduce noise by measuring and maximizing sound levels. This has heightened the tendency of noise-abatement activists and citizens, as chapter 7 showed, to point to measurements in terms of decibels when they try to substantiate the legitimacy of their complaints.[1] Yet the tension between mechanical and pragmatic objectivity reduced, at least in the case of aircraft noise, the actual transparency of how to assess the impact of noise on populations, and thus citizens' control over noise abatement. Moreover, pragmatic objectivity has created sharp boundaries between zones of intervention—boundaries that may contradict citizens' everyday experiences. Owing to such zoning, a citizen's home may be eligible for a state insulation grant, whereas his neighbor's home is not. The resident with the grant may not even be the one most disturbed by noise. And a small change in an aircraft noise index or maximum on the national or European level may suddenly affect airlines' routes, which again will not exactly enhance citizens' feelings of control. Control is even more difficult to acquire in the case of neighborly noise, since it is, as we have seen, left up to citizens themselves to approach their neighbors with the prudence required. In sum, citizens have been made responsible for dealing with the most intangible forms of noise abatement (the ones based on talking others into quieter behavior), and have been distanced from the most tangible ones—the ones based on numbers.

That this has not exactly helped to wipe noise from the public agenda is also due to the fact that feelings of being in or out of control are highly significant for the perception of sound, as both previous chapters and contemporary psychoacoustic research support. The previous chapters have shown that unexpected, unfamiliar, and irregular sounds have often been defined as noise, whereas rhythmical sounds signified situations in which people felt in control. Chapter 2 discussed how such differences are part of deeply rooted cultural repertoires of dramatizing sound. In literature, for instance, the topoi of the intrusive and the sinister sound—with their stress on the chaos and the unexpectedness of sounds, respectively—represent noise, whereas the sensational and the comforting sound stand for its rhythmical opposite. Chapters 3 and 4 have provided ample examples of how such repertoires of dramatizing mechanical sound were linked up or contrasted with definitions of public problems of noise. In addition, chapters 5 and 6 illustrated the increasing cultural significance of controlling the sound of music, both in mass and art music.

Today, psychologists and other social scientists stress that feelings of being in control are the keys to individual well-being and to the prevention of noise nuisance in particular (Hell et al. 1993; Broer 2006; Stallen and van Gunsteren 2002). Acousticians and psychoacousticians therefore plea for stable and predictable aircraft flight paths, or for the direct influence of citizens on the choice of how aircraft pass over their homes, for instance, whether they pass over several times per hour over a short period of time or only a few times per hour over a prolonged period of time (Schreuder 2006; Illusie 2007). Yet, formulas calculating noise impacts have not taken such issues into account. On the contrary, as we have seen in this chapter, the effects of the discontinuity, unfamiliarity, and unexpectedness of sound on the perception of noise have been exactly the issues that have been left out from the standards defining the impact of noise on populations. This occurred when scientists and politicians coupled limits in levels of sound to zoning, known today as the dosis-effect model in noise measurement and abatement.

The *spatial* character of many of the interventions has similarly contributed to the persistence of public problems of noise. Sound crosses borders between neighbors, neighborhoods, and even nations casually, but has often been handled spatially by imposing zones, canalizing traffic, restructuring city planning, and drawing noise maps. We have seen how zoning was the legacy of the long-standing islands-of-silence tradition, and how it, through nuisance law and subsequent legislation, literally became the basis on which other measures came to be piled. Not long after the establishment of the decibel and maximum noise levels, politicians made these levels the twins of zones, which, in the case of aircraft noise, also included night norms. Today, at the core of traffic, transport, and aircraft noise abatement efforts are maps with circles around areas with distinct maxima of allowed noise impact on the population living in these areas.

Cultural geographers have shown how such spatial policies and forms of planning are closely linked to the rise of the nation state, itself geographically defined, and the exertion of social power (Harvey 1990). The example of the insulation programs show, however, how mapping may clash with citizens' conceptions. Even the recently established "silent areas" in the countryside tend to become highly contested spaces, since they embody a promise of sheltering citizens from noise that might lead to deep disappointments in the case of nuisances. Snowmobiling in Yellowstone National Park has, for instance, sparked intense controversy, exactly because many visitors expect protection from noise to go along with nature conservation (Coates 2005: 653).

Maps may also embody the illusion that one can freely move to quieter areas. Both nineteenth-century civil codes and antistreet-music regulations had already treated the wealthier residential quarters as the areas where noise was out of place. Over time, noise has been regionalized and still is "distributed socially unequally" (Ampuja 2005: 86). This phenomenon expresses, as sociologist John Urry claims, how "space makes a clear difference to the degree to which ... the causal powers of social entities [such as class] are realized" (Urry 1995: 13). Islands of silence, such as universities on the one hand, and industrial areas on the other, even "crystallise through spatialisation a separation of Mind from Body" as geographer Doreen Massey stresses. Such knowledge relations are "deeply interwoven with those of class, and the two together are reinforced through spatial form" (Massey 2005: 144). Today's European Union noise maps, mentioned in chapter 7, explicitly aim to influence citizens' choices of where to live. Given the unequal distribution of noise over social class and income, however, such instruments of intervention suggest more freedom than individual citizens actually have, unless they are sufficiently wealthy.

These two issues—the paradox of control and the consequences of space-bound noise abatement measures—have played their parts in keeping noise on the public agenda. Previous chapters have explained the outcomes of subsequent debates of noise on the basis of clashes between dramatizations and definitions of noise, the rise of new cultural practices, such as listening to music while working, discourse coalitions, depoliticization, and pragmatic objectivity. Most of these phenomena contributed to the obduracy of the public problem of noise. Yet the combined results of stacking various forms of noise legislation on another over time have been the paradox of control and the spatial focus of solutions.

STUDYING PUBLIC PROBLEMS IN TECHNOLOGICAL CULTURE

I claimed in chapter 1 that understanding the history of public problems of noise required the use of three domains of study: public-problems theory, science and technology studies (STS), and the interdisciplinary study of the senses. The subsequent chapters have shown how these fields of research have helped me to answer my question about the persistence of noise as a public problem. Public-

problems theory enabled me to unravel the changing structure of public problems of noise and to understand the role of discourse coalitions; STS offered
me several notions of mechanical objectivity as well as insights into the role of
scientists and standardization in liberal democracies, and the field of sensory
studies clarified the symbolism of sound. In this section, I would like to add a
few insights on how to write a sound history of technological culture, in which
"sound" refers to more than the main topic of this book. These insights involve
the masonry of public problems, pragmatic objectivity, and the repertoires of
dramatizing sensory perceptions.

More than anything else, the previous chapters have shown that the sounds
of new technologies that so often took citizens, scientists, and governments by
surprise came to be appropriated and dealt with by the piling up of strategies
of noise abatement on top of another, rather than the replacement of older
forms of tackling noise by completely new ones. The science of acoustics, psychological expertise about the effects of sound on humankind, technologies of
measuring and controlling sound, and the processes of standardization never
did completely change the definitions of noise problems, but instead became
linked to forms of regulation and legislation known from the past. Even the
new notion of measuring sound levels on the basis of the decibel came to be
embedded in the much older notion of zoning.

By saying this, I do not merely claim what STS has shown so frequently:
that technology and culture coconstitute each other. Rather, I aim to stress that
this coevolution often took the form of adding layers—layers that, in turn,
structured the options available later. Long before the late nineteenth century,
our culture had already created zones for blacksmiths and time slots for operating mills. The decibel only added the idea of limiting the level of sound to a
particular maximum in particular areas. This, however, had constraining effects.
It made it more difficult, for instance, to take the effects of the discontinuity
and unexpectedness of sound into account when seeking ways to abate noise. In
technological culture, the general public, politicians, and even scientists themselves often feel caught in technologies' consequences. Yet, the collaborative
work of scientists and politicians in solving public problems is less surprising, but

follows a path of masonry in which one brick is the enabling *and* constraining foundation to another.

Given our Western culture's dependence on many kinds of technologies, and the trust in numbers within our liberal democracies, scientists, engineers, and other academic experts have been and will continue to be asked to give advice on the best solutions to our public problems. A sound history of technological culture should therefore reflect on the role of expertise in political decision making. My contribution has been to show that owing to the tensions between the ideal of mechanical objectivity and the logic of pragmatic objectivity, the solutions proposed by scientists have not always been fully transparent, which has contributed to the paradox of control. My preferred way out of this paradox, however, is not a plea for a full-fledged, direct influence of citizens on the establishment of standards and norms, but for a kind of influence that bears in mind the logic of pragmatic objectivity and the position of expert advisors in technological culture.

As Sheila Jasanoff has recently shown, Western countries with distinct political cultures have, from the 1980s until now, developed different "civic epistemologies," that is, "different understandings of what . . . constitutes adequate citizen participation in technical decision making" (Jasanoff 2005: 271). These differences are expressed in the ways that Western countries establish advisory committees that vary from committees with "instrumentally selected experts" in "contentious" societies like the United States to committees that incorporate "all relevant political viewpoints" in "consensus-seeking" societies like Germany (Jasanoff 2005: 266–267, 270). Our technological cultures cannot do without such ways of enrolling expert advisors. It is useful, however, notably in the case of noise, to enhance citizens' feelings of control by doing two things: having experts design norms that are close to the everyday experiences of citizens (see the last section of this book) and offer choice to citizens *after* general standards have been installed. Pragmatic objectivity necessarily requires strict boundaries that indicate where and how to intervene. Yet even after establishing a maximum noise impact for a particular area, some choice could be left to citizens, as illustrated by the example of choosing between many airplanes flying over in a short period of time versus a few airplanes per hour over a prolonged period of time.

Finally, I would like to stress that a sound history of technological culture should take into account both the cultures of research, as done in chapter 7, and our sensory cultures. Technologies have an impact on societies through the way we perceive them. In preparing for collective action in response to such impacts, we have to inform others about our perceptions by dramatizing our sensory experiences of technology in convincing ways. This is why repertoires of dramatizing perceptions, which in this book were illustrated by the perceptions of mechanical sound, are so crucial for understanding the outcome of debates about public problems in technological cultures. And as it has become clear how such repertoires draw on a long-standing symbolism of sound, I hope to have illustrated how a sound history of technological culture can be enriched by studying the history and anthropology of the senses.

Ecological Modernization, Sensibility in Law, and Complaining in Style

Let me conclude with three alternative approaches to noise abatement: ecological modernization, sensibility in law, and complaining in style. Although this book treats the history of noise largely as a tragedy, its occasional successes *and* its structural aspects open up some alternative ideas for tackling noise.

As I have stressed, there is no need to believe that economic progress and environmental concerns can never be coupled. The history of industrial and aircraft noise provided us with examples of when it is not. Nor is it necessary, contrary to what 1970s legislation suggested, to wait passively for technological innovations in noise reduction. So far, standards such as those for aircraft noise have largely been employed in a *defensive* manner. Sheltering homes from noise through sound insulation is an example of a defensive approach. The Dutch physicist Guus Berkhout has suggested an alternative, which in my view employs an *offensive* approach. Indexes, Berkhout claims, should be linked to goals in noise reduction in order to invite innovation by those involved in the noise problem, such as the aircraft industry. In suggesting this strategy, he implicitly adheres to "ecological modernization." Ecological modernization accepts rather than rejects modernization, considers science and technology both as the causes of and the solutions to environmental problems, and takes market dynamics and economic agents into account (Misa et al. 2003).

A second alternative to mainstream noise abatement would be to introduce sensibility in the law in such a way that standards are closer to the everyday experiences of citizens. Instead of the highly abstract standards now in use, one could, for instance, link the maximum movements and flight paths of aircraft to how often people, on average, wake up during the night because of aircraft noise: would being awakened three times a night be acceptable?[2] In addition, it could be helpful to rethink the most long-standing elements of noise abatement once in a while. We know that the special protection of churches, schools, and hospitals has roots in our history of religion, science, and medicine. But today is it still helpful to distinguish between the activities in universities and residential housing as we did centuries ago?

What anyone seeking attention for problems of noise could do as a third alternative is to complain in style. Complaining elegantly was the strategy of the columnists and journalists whose writings I quoted in the first chapter of this book. Many of them complained of noise in a manner full of humor and rhythm, thus making sure that they could raise the issue that bothered them while keeping two back doors open for escape in case the reader would identify them as bores. The first back door was that of irony, the second that of admiration for the virtuosity of their style. The first strategy is a tricky one for those serious about noise, but the second one is highly advisable. Some of topoi of dramatizing mechanical sound discussed in chapter 2, as well as the literary instruments with which it is connected, could be a source of inspiration. Complainants should bear in mind, though, that complaining in style also requires employing dramatizations of mechanical sound that match the definitions of noise prevailing in the setting in which the complainants aim to intervene, or that enable discourse coalitions with others. If they don't attune the style of their complaints to the contexts they aim to change, they will usually not be heard.

If nothing helps, one could follow Doctor Murke—the hero of a short story written by Heinrich Böll in the late 1950s. Murke, a psychologist, works at a radio station. It is his responsibility to edit radio programs to the precise length required for airing. He cuts out moments of silence, collects the remaining band pieces, and sticks them together. Every evening when off work, he listens to the recorded seconds, his anthology of silence. Employees are not supposed to take recordings home. Yet the moment Murke's chief finds out about the snippets, Murke merely regrets one thing: that his collection encompasses no more than three minutes of sampled silence. He does not blame himself, however, for "we do not fall silent very often, do we?"—"*es wird ja auch nicht viel geschwiegen*" (Böll 1987/1958: 45).

Notes

Chapter One

1. Archives World Soundscape Project, Vancouver, Article Files, File "Noise-Historical." The document is based on an analysis of the *Readers Guide to Periodical Literature*.

2. For instance, see the website of Citizens for the Abatement of Aircraft Noise at http://www .caan.org/othergrp.html (March 12, 2007) and The UK Quiet Pages at http://www.quiet.org .uk (March 12, 2007).

3. In Britain, the UK Noise Association is the umbrella organization for such pressure groups (Martin 2007: 5). Similarly, in the United States, the Noise Pollution Clearinghouse started Quietnet for noise abatement groups that do not have their own websites. Usually organizations involved in environmental protection, such as the National Society for Clean Air and Environmental Protection (NSCA) in the United Kingdom, are engaged with noise as well.

4. See the website of the Acoustical Society of America at http://asa.aip.org/links.html (March 12, 2007) and Noisenet, a website with a commercial directory for acoustical products and services at http://www.noisenet.org (March 12, 2007). Both of these have their own international umbrella organizations, the International Commission on Acoustics and the European Acoustics Association, respectively.

5. See *European Commission Green Paper on Future Noise Policy* (COM/96/0540), Executive Summary, http://europa.eu/documents/comm/green_papers/index_nl.htm#1996 (June 14, 2007).

6. See http://interact.uoregon.edu/MediaLit/WFAE/Home (March 13, 2007).

7. For this view, see also the interview the author had with Hildegard Westerkamp, Vancouver, October 26, 1998. According to Westerkamp, the historical aspect of the project had no sufficient intent behind it.

8. Proceedings of the Dutch Parliament of the Lower House, 1975–1976, 13639, no. 1–4.

CHAPTER TWO

1. The *Anti-Noise League* was established in 1934 and changed its name to *Noise Abatement League* in 1937 (Editorial 1938a: 3). Its magazine, *Quiet,* started in 1936 and initially appeared three times a year.

2. Elsewhere, however, Smith mentions seventy dB as the maximum level of sound in early modern England (Smith 1999: 58).

3. Archives World Soundscape, Vancouver (hereafter Archives WSP), interview with Bruce Davis and Barry Truax by Ingemar von Heijne, November 3, 1975, 2.

4. Festival Program, *Musica Sacra Maastricht,* September 18, 2004, p. 3.

5. For instance, Rodney Needham claims that there is profound evidence for the connection between the noise of percussion and transition. He argues that percussion which evokes noise is a widely employed marker of transition or *rites de passage,* such as birth, initiation, marriage, sacrifice, lunar rites, declaration of war, or the reception of strangers (Needham 1967).

6. For an analysis of the manifestations and meanings of "dirt" in a materials science laboratory, see Mody (2001).

7. In the early 1970s, the American doctoral candidate in Speech Sciences Thomas J. Bruneau distinguished "psycholinguistic silences" in speech from "interactive silences" in dialogue and from "sociocultural silences." Sociocultural silences are characteristic ways "in which entire social and cultural orders refrain from speech and manipulate both psycholinguistic and interactive silences" (Bruneau quoted in Tannen and Saville-Troike 1985: 236). Here I focus on sociocultural silences.

8. It is worth noting that the first prescriptions for maintaining silence in monastery scriptoria appeared as early as in the ninth century, whereas reading in silence in contrast to reading

aloud, became common in the tenth century. The reformation's emphasis on the right of individuals to read the word of God without the mediation of a priest strengthened this trend (Manguel 1999/1996: 57–70).

9. The term absolute music, which is derived from the Latin word *absolutus,* meaning "free of," refers to instrumental music that is devoid of any implicit or explicit references to extramusical reality, such as the sound of birds (Randel 2003: 1).

10. The selection process that WSP members used to collect the quotations was not systematic. They collected quotations from novels with which they were already acquainted and asked students of Simon Fraser University (which housed the WSP project) to submit quotations. Each quotation was transcribed onto a card. However, many were recorded on more than one card because they could be ascribed to several categories. Moreover, a few novels, apparently rich in descriptions on sound, were cited many times. See Archives WSP, WSP Project Definitions, Subproject 1: Glossary of Sounds in Literature.

11. The categories of sound included urban soundscapes, domestic soundscapes, factories and offices, entertainment, mechanical sounds, indicators, silence, quiet, echo and reverberation in urban and building environments, reporter's attitude to sound, Western sound as magic, and sound symbolism.

CHAPTER THREE

1. The Statute of the Streets during the reign of Queen Elizabeth I (1558–1603) is well known. It prescribed that "no man shall blow any horn in the night, or whistle after the hour of nine o'clock in the night under pain of imprisonment." A person was also not allowed to make any "sudden outcry in the still of the night, or making any affray, or beating his wife" (Warwick 1968: 64). Yet the Bellman who had to summon—at first by crying, and later by ringing a bell—and check the proper execution of the order of lights at dark *was* permitted to do his work. In 1607, the local authorities of Antwerp prohibited children to play in the vicinity of the Stock Exchange (*Beurs*) because the children were too noisy (Poulussen 1987: 66). Amsterdam had regulations controlling music making on public streets since 1804, and probably even before then (van Dijk 1981: 143, van Daalen 1987: 76–77). In the nineteenth century, the city of Bern had an impressive list of anti-noise regulations, of which the oldest dated back to 1628 (Schafer 1994/1977: 190).

2. The Dutch Supreme Court confirmed this interpretation twice. Archives Nederlandse Stichting Geluidhinder (hereafter NSG), file *Werkgroep Plaatselijke Verordeningen,* no. 2, W.J.K. Brugman, "Notitie inzake jurisprudentie over geluidshinder," March 30, 1973, p. 1.

3. Archives NSG, file *Werkgroep Plaatselijke Verordeningen,* no. 2, W.J.K. Brugman, "Notitie inzake jurisprudentie over geluidshinder," March 30, 1973, pp. 3–4.

4. Originally, nuisances in The Netherlands had been regulated under the Factory Act, which had been preceded by the Factory Decision (1824) and, before that, a Napoleonic decree (1810).

5. Archives WSP, International Bylaws, Germany, Hamburg, *Verordnung zur Bekämpfung gesundheitsgefährdenden Lärms* (May 4, 1965, GVbl. S. 83), and Cologne, *Kölner Strassenordnung,* October 27, 1972, section 10.

6. Philips Concern Archives, File No. 642, Philips Bedrijfsomroep. The year of the report's publication was estimated by Philips archivist, Ivo Blanken.

7. Philips Concern Archives, File No. 642, Philips Bedrijfsomroep, Letter dd. April 23, 1945, Luyten Installatie Bureau E.T.F. Philips Concern Archives, File No. 642, Philips Bedrijfsomroep, and Enar Qwick, "Over muziek bij het werk," p. 2.

8. See the website of the Muzak Corporation at http://www.muzak.com (May 11, 2007).

CHAPTER FOUR

1. Translation quoted from Schafer 1994/1977: 62. See also Bailey 1996, Menninghaus 1996, and Brown et al. 1930. James Sully and Theodor Lessing also mentioned Schopenhauer (see this chapter).

2. For this reason it was "thought indecorous for a corpse to be hustled to the grave in a motor hearse" in the United States until the 1930s (Eliot Morison, quoted in Berger 1979: 183).

3. See also Hogewind 1926: 11, Baron 1982: 169–170, Marwedel 1987: 104–107. The full title of the German journal was *Der Anti-Rüpel (Recht auf Stille) Monatsblätter zum Kampf gegen Lärm, Roheit und Unkultur im deutschen Wirtschafts-, Handels- und Verkehrsleben.* From the second issue onwards, its main title was *Das Recht auf Stille.* In *Der Lärm* (1908a: 43), Lessing also mentioned the Society for the Protection from Street Noise (*Verein zum Schutz gegen den Strassenlärm*) in Nürnberg. However, I have not found any other references to it.

4. Although the New York Society was the "largest and most well-known" antinoise organization of early-twentieth-century America (Smilor 1978: 64), Chicago had its own campaigns at that time, which focused on the noise (whistles, bells, gongs, clappers, horns and cries) of peddlers among other issues (Vaillant 2003; Schultze 1909).

5. The noise of automobiles did not go completely unrestricted, however. Many state laws required exhaust filters (Berger 1979: 103). And from 1908 to 1909, New York City had ordinances that outlawed the excessive use of automobile horns and made mufflers on automobiles obligatory (Smilor 1978: 73).

6. Those who used the older definition followed Hermann von Helmholz. See Sully 1878: 704–706; Pinkenburg 1903: 7; Lessing 1908a: 37–38; Hogewind 1926: 16–18. For the newer type of definition, see Myers 1934: viii; Davis 1937: 6; Dubois 1937: 3. For both definitions, see Fletcher 1929: 99–100.

7. Archives National Physical Laboratory, A.H. Davis, "The measurement of noise," paper read before Section G of the British Association of Aberdeen, September 1934, Advance Proof.

8. The first portabel audiometer was produced in 1924 (Dembe 1996: 182), and the first portable noise meters came into use in 1926 (Free 1930: 24).

9. Kaye worked at the National Physical Laboratory and was also president of the section of Mathematical and Physical Sciences of the British Association for the Advancement of Science (Knudsen 1939: 32).

10. Archives, Sound Foundation (*Geluidstichting*) (1933–1942), abbreviated below as *Archives GS*. At the time I did my research, these archives were housed at the Dutch Acoustical Society (*Nederlands Akoestisch Genootschap*), Delft, The Netherlands. Now, the archives of the Dutch Acoustical Society are in Nieuwegein, The Netherlands. "Stichting voor Materiaalonderzoek, Commissie voor de bestudeering van de acoustische eigenschappen van bouwmaterialen," Document no. 5, 1933.

11. Fokker composed music for a self-designed 31-tone organ (de Bruijn 1984; Rasch 1987).

12. *Archives GS*, "Stichtingsacte Geluidstichting," "Notulen van de vergadering, gehouden op 30 Juli 1934," "Toelichting bij het in het leven roepen van een 'Geluidstichting,'" letter from G.L. Tegelberg to A.D. Fokker (29 October 1934), *Jaarverslag 1935 der Geluidstichting*. See also de Bruijn 1984; Zwikker 1938. Thirty-four institutions participated in the Foundation's Commission of Help, representing local and national government, car drivers, housewives,

tourism, transport, architecture, housing, contracting, engineering, medicine, broadcasting, aircraft, cinema, hotels, restaurants, industry, the Nuisance Act Society (*Hinderwetvereeniging*) and even churches.

13. A similar organization, also established by engineers during this period, was the Heat Foundation (*Warmte-Stichting*). Personal communication with Ernst Homburg, Maastricht, 16 January 2001.

14. *Archives GS:* "Statuten van den Nederlandschen Bond ter Bestryding van Geluidshinder in Nederland," *Jaarverslag van de Geluidstichting over 1937:*2; Jas 1938a; Jas 1938b.

15. *Archives GS:* Pamphlet "Anti-Lawaai Comité Groningen" [1936], "Anti-lawaai-week 23–28 September 1935," "De anti-lawaaimaand te Breda," 2 March 1935, "'Meer Stilte': Anti-lawaai-week in de residentie," *De Nieuwsbron* (19 September 1935), letter from Walter Matthies to Sound Foundation (5 October 1935), letter from "Bond van Bedrijfsautohouders" to Sound Foundation (5 February 1936), *Algemeene* 1936. The campaigns also referred to issues such as the noise of musical instruments and radios. Nevertheless, the control of traffic noise took priority.

16. *Archives GS:* letter from A.D. Fokker to G.L. Tegelberg (27 October 1934); letter from A.D. Fokker to C. Zwikker (28 December 1934).

17. *Archives GS,* "Richtlijnen van den Raad van Bestuur der Geluidstichting nopens plaatselijke bestrijding van geluidshinder," *Annual Reports of the Sound Foundation* (1936, 1937, 1938, and 1940).

18. The Sound Foundation, the KNAC and the Union for Discipline (*Tuchtunie*) collaborated to establish a Permanent Noise Abatement Committee (*Permanent Anti-Lawaai Comité*) in 1947. This committee aimed to abate noise by informing pedestrians, bicyclists, moped riders, and car drivers on traffic regulations. Its activities remained fairly limited, however. See *Tijdschrift voor Politie* (1946–1947), and *De Gong: Orgaan van de Tucht-Unie* (1947–1958).

CHAPTER FIVE

1. The research into the reception of the music and ideas of Russolo, Antheil, and Mondriaan, and into the man-machine debate in music has focused on the journals *Auftakt* (Upbeat), *Caecilia, Internationale Revue i10, Modern Music, Musical Quarterly, Die Musik, Die Musikblätter*

des Anbruch, De Muziek, Revue Musicale, and *De Stijl* (The Style) and on the years between 1910 and 1930. Occassionaly, references to articles in other music journals, such as *Pult und Taktstock* (Desk and Baton), *Melos,* and *Maandblad voor Hedendaagsche Muziek* (Contemporary Music Monthly) have been included.

2. The noise-generating parts of these instruments, built by Russolo with help of his friend Ugo Piatti, were powered by electric motors, handles, or hand bellows, and each sound-generating device produced a characteristic noise. Furthermore, each instrument had a mechanism to amplify sound and control pitch made from chemically prepared drum skin on a frame with a vibrating wire attached to the center of the skin, creating a specific pitch (Brown 1986: 10–11).

Chapter Six

1. This chapter is largely based on K. Bijsterveld (2003). "The City of Din": Decibels, Noise and Neighbors in the Netherlands, 1910–1980, *Osiris 18:* 173–193. Courtesy Chicago University Press.

2. A plate on a telephone box in a post office in the village of Norg in the Netherlands reads, "One should speak clearly, not too fast, and not too loud." The typeface suggests that the plate dates from the 1920s or 1930s. Its message implies that people tended to shout when making telephone calls. See Men 1999, and Schwartz 2003.

3. Officially, residents who felt disturbed by noise from neighbors could also refer to the civil code. Yet, as discussed in chapter 3, the civil code was of no help to citizens who did not own property, and even for those who did, noise was not recognized as a relevant cause of deprivation before 1914.

4. The cities include Rotterdam, The Hague, Leiden, Breda, Utrecht, Amsterdam, Maastricht, and Groningen. They were selected in part on the bases of cross-references in the documentation. City councils faced with problems of gramophone and radio noise often referred to debates or policies in other cities. Other selection criteria include a city's geographical location and size.

5. Municipal Archives (MA) Rotterdam, Printed Papers 1913, No. 180, litt. b., Resolution Draft, Article 1.

6. MA Rotterdam, Printed Papers 1913, No. 180, litt. a., December 29, 1913.

7. MA Rotterdam, Proceedings City Council Rotterdam, Session of February 26, 1914, 64–70, p. 65.

8. MA Rotterdam, Proceedings City Council Rotterdam, Session of February 26, 1914, 64–70, p. 69.

9. MA Rotterdam, Proceedings City Council Rotterdam, Session of February 26, 1914, 64–70, p. 66.

10. MA Rotterdam, Proceedings City Council Rotterdam, Session of February 26, 1914, 64–70, p. 65 and p. 69. The council members used the concepts "phonograph" and "gramophone" interchangeably. The phonograph, patented by Thomas Edison in 1877, recorded sound on a cylinder, whereas the gramophone, patented by Emile Berliner in 1887, had a (gramophone-) disk as sound carrier (de Meyer 1997: 20, 26).

11. MA Rotterdam, Proceedings City Council Rotterdam, Session of February 26, 1914, 64–70, p. 66.

12. MA Rotterdam, Proceedings City Council Rotterdam, Session of February 26, 1914, 64–70, p. 67. For this argument, see also MA Rotterdam, Collection Van Vollenhoven, 1893–1936, 46/61 (November 1913–December 31, 1913, "Verordening wering hinder van muziekinstrumenten").

13. MA Rotterdam, Proceedings City Council Rotterdam, Session of February 26, 1914, 64–70, p. 70.

14. MA Rotterdam, Proceedings City Council Rotterdam, Session April 23, 1914, pp. 177–178.

15. In the Netherlands, the first radio broadcasting station programming for a wide audience began in 1919 (de Goede 1999: 49). For the measures taken against noise created by musical devices, see: MA The Hague, Printed Papers 1927, no. 365 "Wijziging Algemeene Politie-verordening"; MA Leiden, Municipal Publications (*Gemeenteblad*) 1928, no. 21, "Verordening op het maken van Mechanische Muziek en Geluid"; MA Breda, Municipal Publications (*Gemeenteblad)* 1928, no. 458, "Verordening tot wering van hinder van geluidinstrumenten"; MA Utrecht, Municipal Publications (*Gemeenteblad*) 1929, no. 1 "Verordening Straatpolitie, 11de wijziging (art. 56bis)," MA Amsterdam, Archives of the Municipal Police 1814–1956

(1974), Archive no. 5225, Inventory no. 2224, "Herziening Algemeene Politieverordening," art. 67A, 1931; MA Maastricht, Archives City Administration 1851–1969, no. 1.759.42, "Verordening tot wering van hinderlijke radio- en andere mechanische muziek," August 1, 1930; MA Groningen, Appendices Proceedings Municipal Council 1933, no. 203, "Voorstel … tot wijziging van het Reglement van Politie."

16. The proceedings of municipal councils did not mention the party affiliation of the council members who commented on the legislation. Therefore, information about the party background of the relevant council members was obtained from the following sources. For Rotterdam: oral information, Mrs. Van Balen, city archivist, on the basis of an index; The Hague: Maarten van Doorn, ed. (1985). *De Haagse gemeenteraad 1816–1941: Biografische gegevens van haar leden, bijeengebracht door L. Fledderus en H.M.B. Jacobs onder redactie van M. van Doorn*, The Hague: Gemeentearchief 's-Gravenhage; Leiden: oral information, city archivist, and "J.A.H. Manders overleden," *Leidsch Dagblad,* July 10, 1953; Breda, "Lijst van Gemeenteraadsleden/ Leden Stedelijke Raad 1815–1982 (ten dele met politieke richting)," D 593.13678; Amsterdam: P. Hofland (1998). *Leden van de Raad: De Amsterdamse Gemeenteraad 1814–1941.* Amsterdam: Gemeentearchief; Maastricht: MA Maastricht, Inventory no. 352.07.51, no. 12 A "Verkiezingen voor de leden van de Gemeenteraad, 1923–1953" and *Gemeenteverslag* 1927. In Groningen the ordinance was accepted without discussion.

17. MA Breda, Proceedings City Council Breda, Session September 4, 1928: p. 849 and p. 851.

18. MA The Hague, Proceedings City Council The Hague, Session July 11, 1927, 483, and MA Amsterdam, Proceedings City Council Amsterdam, Session October 23, 1930, II: 1987–1988.

19. MA Amsterdam, Proceedings City Council Amsterdam, Session December 10, 1929, II: 2759, II: 2751, II: 2754–2755; and Session October 23, 1930, II: 1993–1994.

20. MA Amsterdam, Proceedings City Council Amsterdam, Session December 10, 1929, II: 2755.

21. MA Maastricht, Archives City Administration 1851–1969, No. 1.759.42, "Verordeningen tot wering van hinderlijk geluid" (1930–1937), letter June 6, 1930, Committee of penal regulations to City Council, and "Verordening tot wering van hinderlijke radio- en andere mechanische muziek," August 1, 1930.

22. Archives *Nederlandse Stichting Geluidshinder,* Delft, File *Werkgroep Plaatselijke verordeningen,* Letter D.J. van Manen to Vereniging Nederlandse Gemeenten, dd. 3 January 1973. The appendix to this letter contains an inventory of the local ordinances on noise from forty-nine Dutch municipalities. Thirty-five of these passed ordinances against the noise nuisance of both musical and mechanical instruments. Only two communities passed ordinances solely prohibiting the noise nuisance of mechanical instruments.

23. MA Amsterdam, Archives Municipal Police 1814–1956 (1974), Archives No. 5225, Inventory No. 5688, letters of complaint, letter May 11, 1936 A.C. van der Bijl Jr. to chief commissioner of police, letter July 16, 1939 W.J. Sjollema; letter July 5, 1937 J. van Beek.

24. MA Leiden, "Verslagen omtrent den toestand van de gemeente Leiden" 1929–1939, MA Breda, "Verslagen van den toestand der gemeente Breda," 1929–1937.

25. MA The Hague, Printed Papers 1927, no. 364, Explanatory Note.

26. MA Amsterdam, Proceedings City Council Amsterdam (Evening) Session October 23, 1930, 2: 2003, MA The Hague, Proceedings City Council The Hague, Session July 11, 1927, 480.

27. MA Breda, Inventory no. 1.759.4, 235/II ("Beperking straatrumoer en hinderlijk lawaai"); letter November 17, 1937, Provincial Executives North Brabant to the Mayors of the Councils of North Brabant.

28. MA Breda, Inventory no. 1.759.4, 235/II, letter January 22, 1940 Provincial Governor North Brabant to the Mayors of North Brabant.

29. MA Breda, Inventory no. 1.759.4, 235/II letter January 27, 1940, deputy Police Chief to city leadership.

30. Bartlett based his claim in part on experiments carried out by Francis L. Harmon. Harmon had claimed that noise effects, such as increases in metabolic, heart, and breathing rates gradually disappeared when the subject was presented "with the same situation day after day, over a period of several weeks" (Harmon 1933: 79).

31. Davis and Kaye 1927, Koelma 1929, Davis 1934, Das lärmfreie 1934a, Das lärmfreie 1934b, Wigge 1936, van Loghem 1936, Dubois 1937.

32. MA Amsterdam, Archive No. 5225 (Amsterdam Police), Inventory No. 5688, File A.8 1937, Notice Police Chief Bakker, February 18, 1937.

33. MA Amsterdam, Archive No. 5225 (Amsterdam Police), Inventory No. 5688, File 1937 "Diversen," "Communiqué."

34. MA Amsterdam, Archive No. 5225 (Amsterdam Police), Inventory No. 5688, File A. 8 1938, 1939, 1940, Police Report April 29, 1938, Notice Police Chief Bakker, August 15, 1938, Notice Police Chief Bakker, June 14, 1939.

35. Archives *Nederlands Akoestisch Genootschap,* Delft, *Loonmeetdienst,* File no. 110, *Silenta,* letter 2 March 1940, H. van Tongeren to C. Zwikker, Letter C. Zwikker to H. van Tongeren [18 March 1940], Letter March 27, 1940 H. van Tongeren to C. Zwikker.

36. MA Amsterdam, Archive No. 5225 (Amsterdam Police), No. 5688, File A. 8 1937, Notice police chief Bakker, dd. 7-8-1937.

37. Normblad V 1070, *Geluidwering in woningen N.B.G. III,* 1951, quoted in Geluidshinder 1961: 10.

38. NEN is the acronym of the Dutch Institute for Normalization (*Nederlands Normalisatie-Instituut*).

39. *Geluidshinder: Rapport van de Gezondheidsraad,* Proceedings Parliament of the Lower House 1971–1972, Printed Documents 11673: 17.

40. Proceedings Parliament of the Lower House, 1975–1976, No. 13639, No. 1–4, p. 131.

41. Proceedings Parliament of the Lower House, 1975–1976, No. 13639, No. 5, p. 12 and p. 4; see also No. 8, p 26.

42. Proceedings Parliament of the Lower House, 1975–1976, No. 13639, No. 6, p. 2; and 1976–1977, no. 13639, no. 9 and 10, p. 2, pp. 50–51.

43. In 1962, the Sound Foundation (*Geluidstichting*) had been transformed into the Dutch Society for Acoustics (*Nederlands Akoestisch Genootschap* (NAG)) in order to strengthen its focus on science. At that moment, the Permanent Noise Abatement Committee, established in 1947 (discussed in chapter 4) still existed formally. Yet from the mid-1950s, its members did not meet anymore, and I have never been able to find its archive. In 1968, the Dutch Congress for Public Health (*Nederlands Congres voor Openbare Gezondheidsregeling*) organized a successful conference on noise that raised a lot of press interest. As a follow-up, the Congress initiated the Dutch Foundation for Noise Abatement. See Archives NAG, Annual Report *Geluidstichting* 1962: 1, 3, Archives NSG, File without title or number, Memorandum "Rol van de voormalige

Geluidstichting en het Nederlands Akoestisch Genootschap bij de bestrijding van de geluid-shinder in Nederland" (1966): 5, 7, Archives NSG, Files "Oprichting NSG," and "Voorberei-ding oprichting NSG," Interview with J. Kuiper, NSG, Delft, November 26 1998.

44. MA Amsterdam, Archives Municipal Police 1814–1956 (1974), Archive no. 5225, Inven-tory no. 5688, Ordonance sur le Bruit, no. 1 1931—Fevrier, Art. 1, City Noise II 1932, Ch. 1, 10–11, Ch. III, 7, Ch. IV, 6–7. See also Rex 1935: 42.

45. See for a similar approach in 1930 Cologne: Rex 1935: 17.

CHAPTER SEVEN

1. Proposal for a Directive of the European Parliament and of the Council relating to the Assessment and Management of Environmental Noise, COM (2000) 468, http://europa.eu .int./comm/environment/docum/00468_en.htm (July, 31, 2007).

2. Proposal (note 1), p.18.

3. The Port of New York Authority (PONYA) governed both the ports and airports of New York City, and had jurisdiction and responsibility independent from the New York City Council.

4. Archives National Physical Laboratory (hereafter Archives NPL), London, Annual Report NPL 1930: 90.

5. Archives NPL, Annual Report NPL 1930: 90; see also Kaye 1931b: 478.

6. Archives NPL, Annual Report NPL 1937: 29.

7. Archives NPL, Annual Report NPL 1937: 29–30.

8. Archives NPL, Papers Robinson, D.W. Robinson [1963], Recent Advances in the Subjec-tive Measurement of Noise, 157–177, p. 164. See also Robinson (1960).

9. Archives NPL, Papers Robinson, D.W. Robinson [1963], Recent Advances in the Sub-jective Measurement of Noise, 157–177, p. 163.

10. Archives NPL, Papers Robinson, D.W. Robinson [1963], Recent Advances in the Subjec-tive Measurement of Noise, 157–177, pp. 164–165.

11. Archives NPL, Annual Reports NPL, 1930–1961.

12. If N=4, then log N = 0,6021 (approx.), so 15 log N = 9 (approx.), which represented the "trade-off."

13. In this regulation, the ICAO categorized all existing and future airplanes into Chapter 1, 2 and 3 airplanes. At the end of the 1970s, the European Civil Aviation Conference (ECAC) decided that after 1987, Chapter 1 airplanes would no longer be permitted (Bouwens and Dierikx 1996: 236).

14. National Archives, The Netherlands (NA), Adviescommissie Geluidhinder door Vliegtuigen (AGV), Inventory no.1, Letter Kosten and Van Os to TPD-TNO and Technical University, [15 January 1962].

15. NA, AGV, Inventory no. 3, "Installatierede," 9 October 1961, Inventory no. 1, "Samenvatting van de bespreking, gehouden op 16 december 1960, inzake een te benoemen commissie voor de geluidshinder door vliegtuigen." Hereafter: "Samenvatting . . .vliegtuigen."

16. NA, AGV, Inventory no.1, "Samenvatting. . . vliegtuigen" (note 15), p. 2; letter C.W. Kosten to the members of the committee on aircraft noise, 16 January 1961, p. 2.

17. NA, AGV, Inventory no. 3, "Installatie en bespreking op 9 oktober 1961," and Inventory no. 1, "Samenvatting . . .vliegtuigen" (note 15), p. 3.

18. NA, AGV, Inventory no. 5, "Interim-verslag over de werkzaamheden tot eind januari 1963," 28 January 1963 (Vliela-Secr. 61), p. 3.

19. NA, AGV, Inventory no. 7, "Concept-rapport aan [het] Ministerie van Verkeer en Waterstaat (Vliela-Secr. 140)," June 1965 [and later], topic no. 104.

20. NA, AGV, Inventory no.1, "Concept-schrijven aan de Directeur-Generaal van de Luchtvaartdienst," Vliela-Secr. 33, [1962], 1–6, Inventory no. 5, "Interim-verslag over de werkzaamheden tot eind januari 1963," 28 January 1963 (Vliela-Secr. 61), 1–5, Inventory no. 5, "The Problem of Annoyance by Aircraft Noise," 7 September 1962 (Vliela-Secr. 43), 1–7.

21. NA, AGV, Inventory no. 1, "Concept-schrijven aan de Directeur-Generaal van de luchtvaartdienst," Vliela-Secr. 33, [1962], p. 3.

22. NA, AGV, Inventory no. 11, "Enkele aantekeningen van voorzitter Prof.dr.ir. C.W. Kosten naar aanleiding van zijn bezoek aan ICAO, Montreal, 10 en 11 maart 1964 en FAA, Washington, 18 en 19 maart 1964," July 1, 1964 (Vliela-Secr. 100), p. 1.

23. NA, AGV, Inventory no. 5, "Report on Mission to International Standards Organization (ISO), Technical Committee 43 (Acoustics) Working Group 12 and 13 in Prague, 19–23 April 1966 . . . by A. Spooner C/OPS." Hereafter: Report Spooner.

24. Report Spooner (note 23), p. 2.

25. Report Spooner (note 23), p. 3 and p. 5.

26. Report Spooner (note 23), p. 1–3.

27. Report Spooner (note 23), p. 4.

28. Report Spooner (note 23), p. 4.

29. See ISO-recommendation R 507, 1st edition (Procedure for Describing Aircraft Noise around an Airport, October 1966). For a detailed explanation of ISO R 507 and its later revision, see Archives TNO-TPD, Papers Van Tol, Memorandum Adviescommissie Geluidshinder door Vliegtuigen, 17 June 1968 (Vliela-secr. 256).

30. Report Spooner (note 23), p. 2.

31. As a rule of thumb, experts expressed the relation between PNdB and dB(A) by adding 12–14 "units" to dB(A) (Bürck 1965: 20). This again underscores the fragility of the compromise between PNdB and dB(A).

32. Archives TNO-TPD, papers Van Tol, Letter ir. G.J. Kleinhoonte van Os to C.A.F. Falkenhagen, 26 June 1969, p. 2.

33. NA, AGV, Inventory no. 12, letter Prof.dr.ir. C.W. Kosten and Ir. G.J. Kleinhoonte van Os to Ir. C.A.F. Falkenhagen, 5 May 1969, p. 6.

34. Archives TNO-PG, Papers Bitter, Letter ir. G.J. van Os to drs. C. Bitter, December 24, 1964.

35. Archives TNO-PG, Papers Bitter, Letter ir. G.J. van Os to drs. C. Bitter, April 26, 1965.

36. Archives TNO-PG, Papers Bitter, Letter prof. dr. ir. C.W. Kosten and ir. G.J. van Os to Drs. C. Bitter, May 25, 1965 (hereafter Letter May 25, 1965), p. 1.

37. Letter May 25, 1965 (note 36), p. 2.

38. Letter May 25, 1965 (note 36), p. 4.

39. Letter May 25, 1965 (note 36), p. 5.

40. Archives TNO-PG, papers Bitter, Letter ir. D. van Zuilen to Ir. G.J. van Os, February 3, 1965.

41. Archives TNO-PG, papers Bitter, Letter Dr. C. Bitter to Ir. G.J. van Os, June 28, 1965.

42. Archives TNO-PG, Papers Bitter, "Voorspelbare reacties van de bevolking op vliegtuig-lawaai in woongebieden, op grond van de in 1963 rond Schiphol op verzoek van de Adviescom-missie Geluidhinder door Vliegtuigen gehouden enquête" (without date, probably second half of 1965), p. 2. See also Bitter 1972: 262–263.

43. NA, AGV, Inventory no. 12, Letter ir. C.A.F. Falkenhagen to Prof.dr.ir. C.W. Kosten, 31 August, 1971, pp. 2–3. See also Bouwens and Dierikx 1996: 274, 277–278.

44. Archives TNO-TPD, papers Van Tol, letter committee on Noise nuisance Norms to Min-ister of Public Health and Environmental Health Care, February 7, 1972, with appendix. See also Bouwens and Dierikx 1996: 330.

45. This foundation emerged from a local committee that had started fighting aircraft noise nuisance in 1962. Archives TNO TPD, "Hoorzitting op 18 Februari 1971 van de Vaste Com-missie van de Tweede Kamer der Staten-Generaal voor Verkeer en Waterstaat, Stiching tegen Geluidshinder door Vliegtuigen," p. 2.

Chapter Eight

1. For this tendency, see the work of the social psychologist Pieter Jan Stallen at http://www .geluidnieuws.nl/2002/feb2002/stallen.html (June 30, 2007).

2. Personal communication with Nicholas Miller, senior vice-president of Harris, Miller, Miller & Hansen Inc., Burlington, Mass., February 12, 2007.

REFERENCES

Aalders, Maria Victor Constantijn (1984). *Industrie, milieu en wetgeving: De Hinderwet tussen symboliek en effectiviteit* Amsterdam: Uitgeverij Kobra.

Aantal (1997). Aantal maandelijkse berichten in de dagbladpers per lawaaibron. *NSG-Nieuws* 20, no. 5:3.

Abbott, Ernest J. (1936). The Role of Acoustical Measurements in Machinery Quieting. *Journal of the Acoustical Society of America* 8 (October): 133–142.

Abrahams, Frits (2000). Met mij. *NRC Handelsblad,* November 14.

Actie (1937). Actie voor meer stilte in het stadsleven. *Het Volk,* November 12, Morning Edition.

Agar, Jon (2002). Bodies, Machines and Noise. In Iwan Rhys Morus, ed., *Bodies/Machines* (197–220). Oxford: Berg.

Agar, Jon (2003). *Constant Touch: A Global History of the Mobile Phone.* Cambridge, U.K.: Icon Books.

Ahrendt, Hannah (1980/1977). *Denken: Deel 1 van 'Het leven van de geest'.* Amsterdam: De Arbeiderspers. (Originally published in 1977 as *Thinking, part 1 of "The Life of the Mind"* [London: Martin Secker & Warburg].)

Aircraft Noise (1972). *Aircraft Noise: Should the Noise and Number Index Be Revised?* London: HMSO/The Noise Advisory Council.

Aisenberg, Andrew R. (1999). *Contagion: Disease, Government, and the "Social Questiom" in Nineteenth-Century France.* Stanford: Stanford University Press.

Alberts, Gerard (2000). Computergeluiden. In Gerard Alberts and Ruud van Dael, eds., *Informatica & Samenleving* (pp. 7–9). Nijmegen: Katholieke Universiteit Nijmegen.

Alberts, Gerard (2003). Een halve eeuw computers in Nederland. *Nieuwe Wiskrant* 22:17–23.

Algemeene (1936). *Algemeene inlichtingen over de lawaai-bestrijding.* Delft: Geluidstichting.

Amphoux, Pascal (1994). Die Zeit der Stille: Urbanität und Sozialität. In Uta Brandes, ed., *Welt auf tönernen Füssen: Die Töne und das Hören* (pp. 86–99). Göttingen: Steidl Verlag.

Ampuja, Outi (2005). Towards an Artificial Soundscape? Modern Soundscapes under Human Design. *ICON* 11:79–94.

Annual, The (1938a). The Annual Dinner. *Quiet* 2, no. 7 (January): 27–32.

Annual, The (1938b). The Annual Dinner. *Quiet* 2, no. 9 (December): 23–24.

Antheil, George (1924). Manifest der musico-mecanico. *De Stijl* 6, no. 8:99–102.

Antheil, George (1925a). Abstraktion und Zeit in der Musik. *De Stijl* 6, no. 10/11:152–156.

Antheil, George (1925b). My Ballet Mécanique. *De Stijl* 6, no. 12:141–144.

Antheil, George (1990/1945). *Bad Boy of Music.* Hollywood: Samual French.

Anti-Lawaai-Campagne (1933). Anti-Lawaai-Campagne van "De Telegraaf": Geluidsmeter zal bewijzen. *De Telegraaf,* October 9, Avondblad.

Antrim, Doron K. (1943). Music in Industry. *Musical Quarterly* 24, no. 3:275–290.

Assael, Brenda (2003). Music in the Air: Noise, Performers and the Contest over the Streets of the Mid-Nineteenth-Century Metropolis. In Tim Hitchcock and Heather Shore, eds., *The Streets of London: From the Great Fire to the Great Exhibition* (pp. 183–197). London: Rivers Oram Press.

Attali, Jacques (1985). *Noise: The Political Economy of Music.* Manchester: Manchester University Press.

Atwood, Margaret (1969). *The Edible Woman.* Toronto: McClelland and Stewart.

Auf (1910). Auf die Mensur! *Der Antirüpel* 2, no. 2:12–13.

Aus (1910). Aus Mittgliederbriefen. *Der Antirüpel* 2, no. 3:20.

Äusserungen (1926). Äusserungen über "Mechanische Musik" in Fach- und Tagesblättern. *Anbruch* 8, no. 8/9:401–404.

Babbage, Charles (1989/1864). Street Nuisances. In Martin Campbell-Kelly, ed., *The Works of Charles Babbage*. Volume 11: *Passages from the Life of a Philosopher.* London: William Pickering. (Originally published as *A Chapter on Street Nuisances*, 1864.)

Bailey, Peter (1996). Breaking the Sound Barrier: A Historian Listens to Noise. *Body & Society* 2, no. 2:49–66.

Bailey, Peter (1998). *Popular Culture and Performance in the Victorian City.* Cambridge: Cambridge University Press.

Ballot, G.H. (1948). Bestrijding van Hinderlijk Lawaai in Fabrieken en Werkplaatsen. *Bedrijf en Techniek* 3, no. 30:54–56.

Bank, Jan, and Buuren, Maarten van (2000). *1900: Hoogtij van burgerlijke cultuur.* The Hague: Sdu Uitgevers.

Barnes, S.J. (1998). *Muzak, the Hidden Messages in Music: A Social Psychology of Culture.* Lewiston: Edwin Mellen Press.

Baron, Lawrence (1982). Noise and Degeneration: Theodor Lessing's Crusade for Quiet. *Journal of Contemporary History* 17:165–178.

Bartlett, Frederic Charles (1934). *The Problem of Noise.* Cambridge: Cambridge University Press.

Bartsch, Ingo, et al. (1986). *Russolo: Die Geräuschkunst, 1913–1931.* Bochum: Museum Bochum Kunstsammlung.

Baudet, Henry (1981). Consumptie in depressie. In J. Hannes, ed., *Consumptiepatronen en prijsindices: Acta van het colloquiium 14 en 15 maart 1980 te Brussel gehouden* (pp. 47–60). Brussels: Vrije Universiteit Brussel.

Beaumont, Antony (1985). *Busoni the Composer.* London: Faber and Faber.

Beck, K., and Holtzmann, F. (1929). *Lärmarbeit und Ohr: Eine klinische und experimentelle Untersuchung.* Berlin: Verlag von Reimar Hobbing.

Becker, Howard S. (1982). *Art Worlds.* Berkeley: University of California Press.

Benjamin, Walter (1980/1927). *Dagboek uit Moskou.* Amsterdam: De Arbeiderspers. (Orginally published in 1927.)

Bell, Alan (1966). *Noise: An Occupational Hazard and Public Nuisance.* Geneva: World Health Organization.

Beranek, L.L., Kryter, K.D., and Miller, L.N. (1959). The Reaction of People to Exterior Aircraft Noise. *Noise Control* 5:23–31.

Berendt, Joachim-Ernst (1985). *Das Dritte Ohr: Vom Hören der Welt.* Reinbek bei Hamburg: Rowohlt Verlag.

Berger, Michael L. (1979). *The Devil Wagon in God's Country: The Automobile and Social Change in Rural America, 1893–1929.* Hamden, Conn.: Archon Books.

Berghaus, Günter (1996). *Futurism and Politics: Between Anarchist Rebellion and Fascist Reaction, 1909–1944.* Providence: Berghahn.

Bergius, R. (1939). Die Ablenkung durch Lärm. *Zeitschrift für Arbeitspsychologie* 12, no. 4:90–114.

Berkhout, Guus (2003). *Dossier Schiphol: Relaas van een falend democratisch proces.* The Hague: A/Z Grafisch Servicebureau.

Berry, Bernard F. (1998). Standards for a Quieter World: Some Acoustical Reflections from the UK National Physical Laboratory. *Noise/News International* 6, no. 2 (June): 74–83.

Best, Joel, ed. (2001). *How Claims Spread: Cross-National Diffusion of Social Problems.* New York: Aldine de Gruyter.

Bestrijding (1998). Bestrijding burenlawaai. *NSG-Nieuws* 21, no. 1:1–3.

Beton (1929). Beton en acoustiek. *Bouwbedrijf* 6, no. 24:476.

Beuttenmüller, Hermann (1908). *Der rechtliche Schutz des Gehörs.* Karlsruhe i.B.: G. Braunschen Hofbuchdruckerei.

Beyer, Robert T. (1999). *Sounds of Our Times: Two Hundred Years of Acoustics.* New York: Springer-Verlag.

Bijker, Wiebe E., Thomas P. Hughes, and Trevor J. Pinch, eds. (1987). *The Social Construction of Technological Systems: New Directions in the Sociology and History of Technology.* Cambridge, Mass.: MIT Press.

Bijsterveld, Karin (2002). Big Brother's Whisper: Muzak, cultural studies en techniekonderzoek. In Jan Baetens and Ginette Verstraete, eds., *Cultural Studies: Een Inleiding* (pp. 99–114). Antwerp: Vantilt.

Birkefeld, Richard, and Martina Jung (1994). *Die Stadt, der Lärm und das Licht: Die Veränderung des öffentlichen Raumes durch Motorisierung und Elektrifizierung.* Seelze (Velber): Kallmeyer.

Bitter, C. (1960). Lawaai en Geestelijke Volksgezondheid. *Maandblad voor de Geestelijke Volksgezondheid* 15:203–214.

Bitter, C. (1967). Geestelijk-hygiënische facetten. *Tijdschrift voor Sociale Geneeskunde* 45:324–329.

Bitter, C. (1972). Geluidhinder door vliegtuigen. In C.J. Lammers, M. Mulder, and M. Schroeder et al., eds., *Menswetenschappen vandaag: twee zijden van de medaille* (pp. 251–268). Meppel: Boom.

Bitter, C., and Horch, Cary (1958). *Geluidhinder en geluidisolatie in de woningbouw II: Sociaal-psychologische aspecten van de geluidhinder.* N.p.: Instituut voor Gezondheidstechniek TNO.

Blankesteijn, Herbert (1998). De geesten in mijn huis. *NRC Handelsblad,* May 20, 36.

Blokland, G.G. van (1958a). Muziek tijdens het werk. *Mens en Onderneming* 12, no. 3:157–176.

Blokland, G.G. van (1958b). *Muziek bij het werk.* Leiden: NITP.

Bois, Yve-Alain, et al. (1994). *Piet Mondriaan: 1872–1944.* Boston: Bulfinch Press.

Böll, Heinrich (1987/1958). *Doktor Murkes gesammeltes Schweigen und andere Satiren.* Cologne: Kiepenheuer & Witsch.

Bonjer, F.H. (1959). "Gevaarlijk geluid": een film. *Tijdschrift voor Sociale Geneeskunde* 37 (November 6): 733–734.

Bourdieu, Pierre (1989). *Opstellen over smaak, habitus en het veldbegrip.* Amsterdam: Van Gennep.

Bouwens, A.M.C.M., and Dierikx, M.L.J. (1996). *Op de drempel van de lucht: tachtig jaar Schiphol.* The Hague: SDU.

Brandes, Uta, ed. (1994). *Welt auf tönernen Füssen: Die Töne und das Hören.* Göttingen: Steidl Verlag.

Braun, Hans-Joachim (1992). Technik im Spiegel der Musik des frühen 20. Jahrhunderts. *Technikgeschichte* 59, no. 2:109–131.

Braun, Hans-Joachim (1994). "I Sing the Body Electric": Der Einfluss von Elekroakustik und Elektronik auf das Musikschaffen im 20. Jahrhundert. *Technikgeschichte* 61, no. 4:353–373.

Braun, Hans-Joachim (1998). Lärmbelastung und Lärmbekämpfung in der Zwischenkriegs-zeit. In Günther Bayerl and Wolfhard Weber, eds., *Sozialgeschichte der Techniek: Ulrich Troitzsch zum 60. Geburtstag* (pp. 251–258). Münster: Waxmann.

Braun, Hans-Joachim, ed. (2002). *Music and Technology in the 20th Century.* Baltimore: Johns Hopkins University Press. (Originally published by Hofheim: Wolke, 2000.)

Breda (1935). Breda, de Anti-Lawaai-Stad: Een onderhoud met Dr. H.J.L. Struijcken. *Dagblad van Noord-Brabant,* April, [n.d.].

Breda bindt (1935). Breda bindt den strijd tegen het stadslawaai aan. *De Telegraaf,* March 30.

Brimblecombe, P., and Bowler, C. (1990). Air Pollution in York 1850–1900. In P. Brimble-combe and C. Pfister, eds., *The Silent Countdown: Essays in European Environmental History* (pp. 182–195). Berlin: Springer Verlag.

Broer, Christian (2006). *Beleid vormt overlast: Hoe beleidsdiscoursen de beleving van geluid bepalen.* Amsterdam: Aksant.

Bronzaft, Arline L. (2002). Noise: The Silent Poison. *Independent,* October 17, 3.

Brown, Barclay (1986). Introduction to Luigi Russolo, *The Art of Noises* (pp. 1–21). New York: Pendragon Press.

Brown, Edward F., et al., eds. (1930). *City Noise: The Report of the Commission Appointed by Dr. Shirley W. Wynne, Commissioner of Health, to Study Noise in New York City and to Develop Means of Abating It.* New York: Noise Abatement Commission, Department of Health.

Brüggemeier, Franz-Josef (1990). The Ruhr Basin 1850–1980: A Case of Large–Scale Envi-ronmental Pollution. In P. Brimblecombe and C. Pfister, eds., *The Silent Countdown. Essays in European Environmental History* (pp. 210–227). Berlin: Springer Verlag.

Bruijn, A. de (1984). *50 jaar akoestiek in Nederland.* Delft: Nederlands Akoestisch Genootschap.

Bruijn, Peter de (2005). Alle snaren trillen. *NRC Handelsblad.* Boeken, May 13, 29.

Brunner (1952). Musik zur Arbeit. *Industrielle Organisation* 21, no. 6:163–167.

Bücher, Karl (1896). *Arbeit und Rhythmus*. Leipzig: S. Hirzel.

Buikhuisen, Wouter (1969). Brozems: een nieuwe lente, een nieuw geluid? *Tijdschrift voor Sociale Geneeskunde* 47, no. 4:7–12.

Bull, Michael (2000). *Sounding Out the City: Personal Stereos and the Management of Everyday Life*. Oxford: Berg.

Bull, Michael, and Les Back, eds. (2003). *The Auditory Culture Reader*. Oxford: Berg.

Bürck, W., et al. (1965). *Fluglärm: Seine Messung und Bewertung, seine Berücksichtigung bei der Siedlungsplanung, Massnamen zu seiner Minderung. Gutachten erstattet im Auftrage des Bundesministers für Gesundheitswesen*. Göttingen: n.p.

Burke, Peter (1993). Notes for a Social History of Silence in Early Modern Europe. In *The Art of Conversation* (pp. 123–141). Ithaca, N.Y.: Cornell University Press.

Burningham, Kate (1998). A Noisy Road or Noisy Resident? A Demonstration of the Utility of Social Constructionism for Analysing Environmental Problems. *Sociological Review* 46:536–563.

Burns, Chester R. (1976). The Nonnaturals: A Paradox in the Western Concept of Health. *Journal of Medicine and Philosophy* 1, no. 3:202–211.

Burns, William (1968). *Noise and Man*. London: John Murray.

Burns, W., and Robinson, D.W. (1970). An Investigation of the Effects of Occupational Noise on Hearing. In G.E.W. Wolstenholme and Julie Knight, eds., *Ciba Foundation Symposium on Sensorineural Hearing Loss* (pp. 177–192). London: J. & A. Churcill.

Burris-Meyer, Harold (1943). Music in Industry. *Mechanical Engineering* 65: 31–34.

Calisch, N.S. (1861). *De Amsterdamsche strafverordeningen: Handboek ter voorkoming van vervolging en veroordeeling wegens politie-overtredingen te Amsterdam*. Amsterdam: M. Schooneveld & Zoon.

Calvesi, M. (1987). *Der Futurismus: Kunst und Leben*. Cologne: Taschen

Campert, Remco (1977). Arbeidsvitaminen. *Volkskrant*, April 16, 1.

Cardinell, R.L. (1943). The Statistical Method in Determining the Effects of Music in Industry. *Journal of the Acoustical Society of America* 15, no. 2 (October): 133–135.

Cardinell, R.L. (1944). *Music in Industry, No. 1–3*. New York: American Society of Composers, Authors and Publishers.

Carson, Anne (1992). *Glass, Irony and God*. New York: New Directions Publishing Corp.

Centraal Bureau voor de Statistiek (1938). *Huishoudrekeningen van 598 gezinnen uit verschillende deelen van Nederland over de perioden 29 juni 1935 t/m 26 juni 1936 en 28 sept. 1935 t/m 25 september 1936, Deel II*. The Hague: Algemeene Landsdrukkerij.

Chadabe, Joel (1997). *Electric Sound: The Past and Promise of Electronic Music*. Saddle River, N.J.: Prentice Hall.

Christ-Brenner, Rudolf (1910). Das Leiden unter dem Lärm. *Der Antirüpel* 2, no. 1:4.

City (1932). *City Noise*, Volume II. New York: Noise Abatement Commission.

Classen, Constance (1993). *Worlds of Sense: Exploring the Senses in History and across Cultures*. London: Routledge.

Classen, Constance (1997). Foundations for an Anthropology of the Senses. *International Social Science Journal* 49 (March): 401–412.

Coates, Peter A. (2005). The Strange Stillness of the Past: Toward an Environmental History of Sound and Noise. *Environmental History* 10, no. 4:636–665.

Cockayne, Emily (2007). *Hubbub—Filth, Noise & Stench in England, 1600–1770*. New Haven: Yale University Press.

Code (1972). *Code of Practice for Reducing the Exposure of Employed Persons to Noise*. London: HSMO.

Coeuroy, André (1920/1921). Les bruiteurs futuristes. *Revue Musicale* 1/2, no. 10:265.

Coeuroy, André (1929). The Esthetics of Contemporary Music. *Musical Quarterly* 15:246–267.

Collier, Richard (1959). *The City That Would not Die: The Bombing of London, May 10–11, 1941*. London: Collins.

Combat (1929). Combat Against Noise, Foreign Letter (Berlin). *Journal of the American Medical Association* 92:2119.

Connell, John (1960). The Story of Noise Abatement. *Quiet Please* 1, no. 1:14–19.

Connell, John (1962). Hon. Secretary's Report. *Quiet Please* 1, no. 3:9.

Contre (1936). Contre le bruit. *Quiet* 1, no. 3, 27.

Control, The (1962). *The Control of Noise: Proceedings of a Conference held at the National Physical Laboratory on 26th, 27th and 28th June, 1961.* London: HMSO.

Conway, Erik M. (2003). Quieting the Jet Engine's Blast: On the Construction and Reduction of "Annoyance," 1950–1980. Paper for Sound Effects: Technology, Practice, and Meaning in the Collection and Evaluation of Sound, *SHOT/4S Annual Meeting,* Atlanta, Georgia, October 16–19.

Copeland, W.C. (1948). A Mobile Laboratory for Acoustical Work. *Journal of Scientific Instruments* 25, no. 3 (March): 82–85.

Copland, A. (1925). George Antheil. *Modern Music* 2, no. 1:26–28.

Corbin, Alain (1986/1982). *The Foul and the Fragant: Odor and the French Social Imagination.* Cambridge, Mass.: Harvard University Press. (Originally published in 1982 as Le miasme et la jonquille: l'odorat et l'imaginaire social, XVIIIe–XIXe siècles [Paris: Aubier Montaigne].)

Corbin, Alain (1995/1991). Time, Desire and Horror: Towards a History of the Senses. Cambridge: Polity Press. (Originally published in 1991 as Le Temps, le Désir et l'Horreur [Paris: Aubier].)

Corbin, Alain (1999/1994). Village Bells: Sound and Meaning in the Nineteenth-Century French Countryside. London: Macmillan. (Originally published in 1994 as Les cloches de la terre. Paysage sonore et culture sensible dans les campagnes au XIXe siècle [Paris: Albin Michel].)

Cowan, Michael (2006). Imagining Modernity through the Ear. *Arcadia* 41, no. 1:124–146.

Cowell, Henry (1930–1931). Music of and for the Records. *Modern Music* 8:32–34.

Cramer, N. (1891). *Van Emden's rechtspraak op de fabriekwet.* The Hague: Belinfante.

Cramer, N. (1904). *De rechtspraak op de Hinderwet.* The Hague: Belinfante.

Daalen, Rineke van (1987). *Klaagbrieven en gemeentelijk ingrijpen: Amsterdam 1865–1920.* Amsterdam: Universiteit van Amsterdam.

Dadson, R.S. (1949). Noise Measurement—A Review of the Problem. In *Noise and Sound Transmission: Report of the 1948 Summer Symposium of the Acoustics Group* (pp. 120–125). London: The Physical Society.

Dam, Willem Jacobus van (1888). *Burengerucht.* 's Hertogenbosch: W.C. van Heusden.

Darby, Ainslie, and Hamilton, C.C. (1930). *England, Ugliness and Noise.* London: P.S. King & Son.

Das lärmfreie (1934a). *Das lärmfreie Wohnhaus.* Berlin: Verein Deutscher Ingenieure.

Das lärmfreie (1934b). Das lärmfreie Wohnhaus. *De Ingenieur* 49, no. 28: B. 104.

Daston, Lorraine, and Galison, Peter (1992). The Image of Objectivity. *Representations* 10, no. 40:81–128.

Davis, A.H. (1931). The Measurement of Noise. In *Report of a Discussion on Audition, Held on June 19, 1931 at the Imperial College of Science* (pp. 82–91). London: The Physical Society.

Davis, A.H. (1934). *Modern Acoustics.* London: G. Bell and Sons.

Davis, A.H. (1937). *Noise.* London: Watts & Co.

Davis, A.H. (1938). An Objective Noise-Meter for the Measurement of Moderate and Loud, Steady and Impulsive Noises. *Journal of the Institution of Electrical Engineers* 83, no. 500:249–260.

Davis, A.H., and G.W.C. Kaye (1927). *The Acoustics of Buildings.* London: G. Bell and Sons.

De Geluidsmeter (1937). De geluidsmeter: Schrik der lawaaiige voertuigen. *De Telegraaf,* May 13, evening edition.

De stiltebrigade (1937). De stiltebrigade op stap. *Het Volk,* November 16, morning edition.

Delft, Dirk van (1995). Een persisterende brom. *NRC Handelsblad,* February 17, 14.

Dembe, Allard E. (1996). *Occupation and Disease: How Social Factors Affect the Conception of Work-Related Disorders.* New Haven: Yale University Press.

DeNora, Tia (2000). *Music in Everyday Life.* Cambridge: Cambridge University Press.

Diederiks, H.A., and Jeurgens, Ch. (1988). Milieu, bedrijf en overheid: Leiden in de negentiende eeuw. *Jaarboek voor de Geschiedenis van Bedrijf en Techniek* 5:427–445.

Dierikx, M., and Bouwens, Bram (1997). *Building Castles of the Air: Schiphol Amsterdam and the Development of Airport Infrastructure in Europe, 1916–1996.* The Hague: SDU Publishers.

Dietz, Egon (1995). De prijs van lawaai. *Index: Feiten en cijfers over onze samenleving 2,* no. 6 (January): 20–21.

Dijk, Marie van (1981). Het beleid van de Amsterdamse overheid ten aanzien van straatmuzikanten 1900–1980. *Volkskundig Bulletin* 7, no. 2:143–158.

Dijk, M. van (1996). Het schildersdoek van de muziek: Ballet mécanique van Fernand Léger en George Antheil. *Mens en Melodie* 51, no. 2: 58–62.

Dillenburger, H. (1957). *Das praktische Autobuch*. Gütersloh: G. Bertelsmann Verlag.

Dincklage, E. von (1979). Ueber den Lärm. *Nordwest* 2:413–414.

Diserens, Charles M. (1926). *The Influence of Music on Behavior*. Princeton: Princeton University Press.

Douglas, Mary (1976/1966). *Reinheid en gevaar*. Utrecht: Het Spectrum. (Originally published in 1966 as *Purity and Danger* [London: Routledge & Kegan Paul].)

Douglas, Susan J. (1999). *Listening In: Radio and the American Imagination from Amos 'n' Andy and Edward R. Morrow to Wolfman Jack and Howard Stern*. New York: Times Books.

Doves, Tamara (2004). Een hele dag zonder lawaai tot op het toilet. *Volkskrant,* 28 April.

Dubois, A. (1937). *Lawaai en lawaaibestrijding*. The Hague: Hinderwetvereeniging/ Geluidstichting.

Dubois, A. (1938). Voorwoord van den Voorzitter. *Stilte* 1:1, 3.

Editorial Notes (1936). Editorial Notes. *Quiet* 1, no. 2, June, 3–5.

Editorial Notes (1938). Editorial Notes. *Quiet* 2, no. 8, April, 3–7.

Editorial Notes (1939). Editorial Notes. *Quiet* 2, no. 10, 3–7.

Eerste aanval (1933). Eerste aanval in de binnenstad. *De Telegraaf,* October 10, Evening Paper.

Effecten (1997). *Effecten van voorlichting inzake burenlawaai*. Delft: Nederlandse Stichting Geluidhinder.

Eijk, J. van den (1960). Voorkómen van geluidhinder in woningen: Wat is wenselijk? Wat is mogelijk? *Maandblad voor de Geestelijke Volksgezondheid* 15:277–286.

Eijk, J. van den (1969). Geluidhinder en woning. *Tijdschrift voor Sociale Geneeskunde* 47, no. 4:18–23.

Egyedi, Tineke M. (1996). *Shaping Standardization: A Study of Standards Processes and Standards Policies in the Field of Telematic Services*. Delft: Technische Universiteit Delft.

Erlmann, Veit, ed. (2004). *Hearing Cultures: Essays on Sound, Listening and Modernity.* Oxford: Berg.

Felber, Erwin (1926). Step-children of Music. *Modern Music* 4, no. 1:31–33.

Feld, Steven (2003). A Rainforest Acoustemology. In Michael Bull and Les Back, eds. (2003). *The Auditory Culture Reader* (pp. 223–239). Oxford: Berg.

Fletcher, Harvey (1929). *Speech and Hearing.* London: Macmillan and Co.

Folkerth, Wes (2002). *The Sound of Shakespeare.* London: Routledge.

Fraunholz, Uwe (2003). Polizei und Automobilverkehr in Kaiserreich und Weimarer Republik. *Technikgeschichte* 70, no. 2:103–134.

Free, Edward Elway (1930). Practical Methods of Noise Measurement. *Journal of the Acoustical Society of America 2,* no. 1:18–29.

Fucini, Renato (1913). *Napoli a occhio nudo: Lettere ad un amico.* Florence: Bemporad.

Galt, Rogers H. (1930). Results of Noise Surveys, Part I: Noise out of Doors. *Journal of the Acoustical Society of America 2,* no. 1:30–58.

Ganzevoort, A.W. (1955). *De auto en zijn baas.* The Hague: Uitgeverij W.van Hoeve.

Garner, J.F. (1975). *Control of Pollution Act 1974.* London: Butterworth.

Gartman, David (1994). *A Social History of American Automobile Design.* London: Routledge.

Gatty, Nicholas C. (1916). Futurism: A Series of Negatives. *Musical Quarterly* 2, no. 1:9–12.

Geluidhinder (1965). *Geluidhinder en geluidwering in de woningbouw.* Rotterdam: Bouwcentrum.

Geluidhinder (1971). *Geluidhinder in het woonmilieu.* Delft: Nederlandse Stichting Geluidhinder.

Geluidhinder [1992]. *Geluidhinder en wat we er aan kunnen doen.* Delft: Nederlandse Stichting Geluidhinder.

Geluidshinder (1961). *Geluidshinder in de woning.* Delft: Geluidstichting.

Geluidsisolatie (1961). Geluidsisolatie tussen woningen. *Tijdschrift voor Sociale Geneeskunde* 39 (March 17): 212.

Geluidsopmetingen (1933). Geluidsopmetingen in andere groote steden. *De Telegraaf,* Avondblad, November 2.

Geluidsoverdaad (1937). Geluidsoverdaad overal in de stad. *Het Volk,* November 16, evening edition.

Giele, Jacques (1981). *Een kwaad leven: Heruitgave van de 'Enquête betreffende werking en uitbreiding der wet van 19 September 1874 (Staatsblad No. 130) en naar den toestand van fabrieken en werkplaatsen' (Sneek, 1887), bezorgd en ingeleid door Jacques Giele. Deel I Amsterdam.* Nijmegen: Link.

Glinsky, Albert (2000). *Theremin: Ether Music and Espionage.* Champaign: University of Illinois Press.

Goede, P. de (1999). *Omroepbeleid met en tegen de tijd: Interacties en instituties in het Nederlandse omroepbestel.* Amsterdam: Otto Cramwinckel.

Goldmark, Josephine (1913). *Fatigue and Efficiency: A Study in Industry.* New York: Survey Associates.

Goodwin, Jeff, and Jasper, James M. (2003). *The Social Movements Reader: Cases and Concepts.* Malden, Mass.: Blackwell Publishing.

Gorky, Maxim (1952/1925). *The Artamonovs.* Moscow: Foreign Language Publishing House. (Originally published in 1925 as *Delo Artamonovych* [Berlin: Kniga].)

Göttgens, Caroline (2005). Muzikale slakken en zingende zwanen. *Dagblad de Limburger,* July 30, C2.

Government (1962). H.M. Government Committee on the Problem of Noise. *Quiet Please* 1, no. 3:15–18.

Graaf, P. de, and H. de Rétrécy (1961). *Wij en onze auto.* Zwolle: La Rivière & Voorhoeve. (Originally published as Arthur Westrup, *Im Auto Zuhause* [Bielefeld: Verlag Delius, Klasing & Co.].)

Groen, J. (1948). Werk en werkomgeving: Muziek, *Tijdschrift voor interne bedrijfsorganisatie* 3:144–148.

Groot, Rokus de (1981). *Achtergrondmuziek.* The Hague: Haags Gemeentemuseum.

Gusfield, Joseph R. (1981). *The Culture of Public Problems: Drinking-Driving and the Symbolic Order.* Chicago: University of Chicago Press.

Güttich, H. (1965). Gewerblich bedingte Lärmschäden. *Münchener Medizinische Wochenschrift* 107, no. 29 (July 16): 1397–1406.

Haahs, Hans (1927). Ueber das Wesen mechanischer Klaviermusik. *Anbruch* 9:351–53.

Haberlandt, Michael (1900). Vom Lärm. In *Cultur im Alltag: Gesammelte Aufsätze von Michael Haberlandt* (pp. 177–183). Vienna: Wiener Verlag.

Hajduk, John C. (2003). Tin Pan Alley on the March: Popular Music, Word War II, and the Quest for a Great War Song. *Popular Music and Society* 26, no. 4 (December): 497–512.

Hajer, Maarten A. (1995). *The Politics of Environmental Discourse: Ecological Modernization and the Policy Process.* Oxford: Oxford University Press.

Halpin, D.D. (1943). Industrial Music and Morale. *Journal of the Acoustical Society of America* 15, no. 2:116–123.

Hanlo, Jan (1958/1946). Nooit meer stil. In *Verzamelde gedichten.* Amsterdam: G.A. van Oorschot. (Originally published in 1946.)

Harmon, Francis L. (1933). *The Effects of Noise upon Certain Psychological and Physiological Processes.* New York: Archives of Psychology.

Harvey, David (1990). *The Condition of Postmodernity.* Oxford: Blackwell Publishers.

Häsker, H. (1910). Für Deutschland. *Der Antirüpel* 2, no. 2 (February): 11.

Heinsheimer, Hans (1926). Kontra und Pro. *Anbruch* 8:353–356.

Heinsius J., ed. (1916). *Woordenboek der Nederlandsche Taal, Deel 8, eerste stuk kr-lichamelijk.* The Hague.

Hek, Youp van 't (2000). Ochtendmens. *NRC Handelsblad,* July 29.

Hell, Jürgen, et al. (1993). Geluid, lawaai en psychopathologie. *Dth: kwartaalschrift voor directieve therapie en hypnose* 13, no. 3:238–258.

Hellpach, Willy (1902). *Nervösität und Kultur.* Berlin: Verlag von Johannes Räde.

Hesse, Hermann (1969/1921). *Steppenwolf.* New York: Bantam Books. (Originally published in 1921 as *Der Steppenwolf* [Berlin: Fischer].)

Het groote-stadsleven (1933). Het groote-stadsleven eischt wellevendheid. *De Telegraaf,* December 22, evening edition.

Hogewind, Frederik (1926). *Analyse en meting van het dagrumoer.* Utrecht: L.E. Bosch & Zoon.

Holl, K. (1926). Musik und Maschine. *Der Auftakt* 6, no. 8:173–177.

Hoove, Sanne ten (2005). Vliegtuiglawaai Schiphol beter meten. *De Volkskrant,* June 9, 3.

Horder, Lord (1936a). Foreword to *Quiet* 1, no. 1, 5.

Horder, Lord (1936b). The Sports Car, the Law and the Public: Is It an Impasse? *Quiet* 1, no. 3, Autumn, 7.

Horder, Lord (1937). Noise and Health. *Quiet* 1, no. 5 (July): 10–11, 15.

Hughes, Thomas P. (1989). *American Genesis: A Century of Invention and Technological Enthusiasm 1870–1970.* New York: Penguin Books.

Husch, J.A. (1984). *Music of the Workplace: A Study of Muzak Culture.* PhD dissertation, University of Massachusets.

Idhe, Don (1976). *Listening and Voice: A Phenomenology of Sound.* Athens: Ohio University Press.

Illusie (2007). Illusie van invloed vermindert hinder. *NRC Handelsblad,* May 1, 8.

Invloed (1960). *De invloed van lawaai op groepen.* Leiden: Instituut voor Gezondheidstechniek TNO.

Jackson, Anthony (1968). Sound and Ritual. *Man: A Monthly Record of Anthropological Science* 3:293–299.

Jarass, Hans D. (1983). *Bundes-Immissionsgesetz: Kommentar.* Munich: C.H. Beck.

Järviluoma, Helmi, ed. (1994). *Soundscapes: Essays on Vroom and Moo.* Tampere: Tampere University.

Jas, W.L.G. (1938a). De oprichting van den Anti-Lawaaibond en een overzicht van het eerste werkjaar. *Stilte* 1, no. 1:4–9.

Jas, W.L.G. (1938b). Overzicht van hetgeen reeds op het gebied der geluidsbestrijding in Nederland werd verricht. *Stilte* 1, no. 1:10–14.

Jasanoff, Sheila (2005). *Designs on Nature: Science and Democracy in Europe and the United States.* Princeton: Princeton University Press.

Jasanoff, Sheila, and Wynne, Brian (1998). Science and Decisionmaking. In Steve Rayner and Elisabeth L.Malone, eds., *Human Choice and Climate Change*. Volume 1: *The Societal Framework* (pp. 1–87). Columbus, Ohio: Battelle Press.

Jemnitz, Alexander (1926). Antiphonie. *Anbruch* 8, no. 8/9:350–353.

Johnson, James H. (1995). *Listening in Paris*. Berkeley: University of California Press.

Jong, R.G. de (1990). Review of Research Developments in Community Response to Noise. In Birgitta Berglund and Thomas Lindvall, eds., *Noise as a Public Health Problem: New Advances in Noise Research, Part II* (pp. 99–113). Stockholm: Swedish Council for Building Research.

Jong, R.G. de, Steenbekkers J.H.M., and Vos, H. (2000). *Hinder en andere zelf-gerapporteerde effecten van milieuverontreiniging in Nederland*. Leiden: TNO.

Jung, Emil (1902). *Radfahrseuche und Automobilen-Unfung: Ein Beitrag zum Recht auf Ruhe*. Munich: August Schupp.

Kahn, Douglas, ed. (1992). *Wireless Imagination: Sound, Radio, and the Avant-Garde*. Cambridge, Mass.: MIT Press.

Kahn, Douglas (1999). *Noise, Water, Meat: A History of Sound in the Arts*. Cambridge, Mass.: MIT Press.

Kaye, G.W.C. (1931a). Noise and Its Measurement. *Supplement to Nature* 128, no. 3224 (August 15): 253–264.

Kaye, G.W.C. (1931b). The Measurement of Noise. *Proceedings of the Royal Institution of Great Britain* 26:435–488.

Kaye, G.W.C. (1937). Noise and the Nation. In *Report of the Meeting of the British Association for the Advancement of Science* (pp. 25–56). N.p.: n.p.

Kaye, G.W.C., and Dadson, R.S. (1939). *Noise Measurement and Analysis in Relation to Motor Vehicles*. Reprinted from Vol. 33 of the Proceedings of the Institution. London: The Institution of Automobile Engineers.

Kerse, C.S. (1975). *Noise*. London: Oyez Publishing.

Kingsbury, B.A. (1927). A Direct Comparison of the Loudness of Pure Tones. *Physical Review* 29 (April): 588–600.

Klosterkötter, W. (1974). Lärmforschungsprojekte in der Bundesrepublik Deutschland unter besonderer Berücksichtigung eigener Arbeiten. In Bruno Böhlen, ed., *Lärmbekämpfung. Refer-*

ate des 8. Internationalen Kongresses für Lärmbekämpfung der AICB, Association Internationale contre le Bruit, im Rahmen der 6. Fachtagung PRO AQUA—PRO VITA 1974 in Basel (pp. 9–27). Zurich: BAG Brunner Verlag.

Knudsen, Vern O. (1939). An Ear to the Future. *Journal of the Acoustical Society of America* 11 (July): 29–36.

Koelma, M. (1929). Meer stilte. *Bouwbedrijf* 6, no. 17:336–340.

Korczynski, Marek, and Jones, Keith (2006). Instrumental Music? The Social Origins of Broadcast Music in British Factories. *Popular Music* 25, no. 2:145–164.

Kosten, C.W. (1967). *Geluidhinder door vliegtuigen.* Delft: Adviescommissie Geluidhinder door Vliegtuigen.

Kosten, C.W. (1969). Het geluidshinderprobleem, zoals de acousticus het ziet. *Tijdschrift voor Sociale Geneeskunde* 47, no. 4:34–40.

Kowar, Helmut (1996). *Mechanische Musik: Ein Bibliographie und eine Einführung in systematische und kulturhistorische Aspekte Mechanischer Musikinstrumente.* Wien: Vom Pasqualatihaus.

Krenek, Ernst (1927). Mechanisierung der Künste. *Internationale Revue* 10, no. 1:376–380.

Krömer, Siegfried (1981). *Lärm als medizinisches Problem im 19. Jahrhundert.* Mainz: Johannes Gutenberg-Universität.

Kruithof, C.L. (1992). Rituele mobilisering: Een kanttekening over gedragsbeïnvloeding. *Tijdschrift voor Sociale Wetenschappen* 37, no. 1:44–52.

Ku, Ja Hyon (2006). British Acoustics and Its Transformations from the 1860s to the 1910s. *Annals of Science* 63, no. 4:395–424.

Kuhn, Thomas S. (1970/1962). *The Structure of Scientific Revolutions.* Chicago: University of Chicago Press. (Originally published in 1962.)

Kuiper, J.P. (1972). Veiligheidswet en lawaaibestrijding. *Tijdschrift voor Sociale Geneeskunde* 50, no. 9, Supplement 2: 42–43, 45.

Kurtze, Günther (1964). *Physik und Technik der Lärmbekämpfung.* Karlsruhe: Verlag G. Braun.

Laan, Mannus van der (1999). Stilte. *Folia*, October 15.

Labrie, Arnold (1994). *Het verlangen naar zuiverheid: Een essay over Duitsland.* Maastricht: Rijksuniversiteit Limburg.

Lagendijk, Ad (1998). Geluidsbarrière. *De Volkskrant,* November 14.

Laird, Donald A. (1930). The Effects of Noise. *Journal of the Acoustical Society of America* 1, no. 2:256–262.

Lammers, B. (1964). Gehoorbescherming in de textielindustrie. *Tijdschrift voor Sociale Geneeskunde* 42:248–252.

Lange, P.A. de, and Janssen, J.H. (1973). To Professor Dr. Ir. C.W. Kosten on his 60th birthday. *Acustica* 28, no. 1:1–2.

Lanza, Joseph (1994). *Elevator Music: A Surreal History of Muzak, Easy-Listening, and Other Moodsong.* New York: Picador.

Latour, Bruno (1987). *Science in Action: How to Follow Scientists and Engineers through Society.* Milton Keynes: Open University Press.

Law, The (1969). *The Law on Noise.* London: The Noise Abatement Society.

Lawaai (1958). Lawaai-en beroepsdoofheid. *Tijdschrift voor Sociale Geneeskunde* 36 (December 5): 643–646.

Lawaaibestrijding (1959). Lawaaibestrijding. *Tijdschrift voor Sociale Geneeskunde* 37 (November 6): 740.

Lawaaimachine (1999). Lawaaimachine. *De Volkskrant,* October 18.

Lawrence, D.H. (1915). *The Rainbow.* New York: Viking Press.

Lee, Margaret M. (1936). English Pioneers of Noise Abatement. *Quiet* 1, no. 2 (June). 18–19.

Lees, Loretta (2004). Urban Geography: Discourse Analysis and Urban Research. *Progress in Human Geography* 28, no. 1:101–107.

Leeuwen, H.A. van (1958). Bedrijfsgeneeskundige aspecten van de beroepsslechthorendheid. *Tijdschrift voor Sociale Geneeskunde* 36 (December 5): 623–631.

Lehmann, Gunther (1961). *Die Einwirkung des Lärms auf den Menschen.* Cologne: Westdeutscher Verlag, Arbeitsgemeinschaft für Forschung des Landes Nordrhein-Westfalen, vol. 94.

Lehmann, Gunther (1964). Conference d'introduction. In *Troisième Congrès International pour la lutte contre le bruit, Paris 13–15 Mai 1964* (pp. 10–14). Paris: Association Internationale Contre le Bruit.

Lemaire, Ton (1995). Het recht op stilte. *De Gids* 158, no. 2:106–108.

Lentz, Matthias (1994). "Ruhe ist die erste Bürgerpflicht": Lärm, Großstadt und Nervosität im Spiegel Theodor Lessings' Antilärmverein. *Medizin, Gesellschaft und Geschichte* 13:81–105.

Lentz, Matthias (1998). Eine Philosophie der Tat, eine Tat der Philosophie. Theodor Lessings Kampf gegen den Lärm. *Zeitschrift für Religions-und Geistesgeschichte* 50, no. 3:242–264.

Lessing, Theodor (1908a). *Der Lärm, Eine Kampschrift gegen die Geräusche unseres Lebens.* Wiesbaden: Verlag von J.F. Bergmann.

Lessing, Theodor (1908b). Die Lärmschutzbewegung. *Dokumente des Fortschritts* 1 (October): 954–961.

Lessing, Theodor (1909). Ueber Psychologie des Lärms. *Zeitschrift für Psychotherapie und medizinische Psychologie* 1:77–87.

Lessing, Theodor (1911a). Unsere Neuorganisation. *Der Antirüpel* 3, no. 4 (April):17–18.

Lessing, Theodor (1911b). Zur Neuorganisation des Antilärmvereins. *Der Antirüpel* 3, no. 6 (June):25–26.

Levin, Miriam R. (2000). *Cultures of Control.* Amsterdam: Harwood Academic Publishers.

Libin, Laurence (2000). Progress, Adaptation, and the Evolution of Musical Instruments. *Journal of the American Musical Instrument Society* 26:187–213.

Lindeman, H.E., and van Leeuwen, P. (1973). Spraakverstaan en gehoorbeschermers. *Tijdschrift voor Sociale Geneeskunde* 51:331–335, 347.

Loghem, J.B. van (1936). *Acoustisch en thermisch bouwen voor de praktijk.* Amsterdam: L.J. Veen's Uitgeversmaatschappij N.V.

Longer (1937). Longer Rails and Quieter Travel. *Quiet* 1, no. 6 (October): 25–26.

MacLeod, Bruce (1979). Facing the Muzak. *Popular Music and Society* 7, no. 1:18–31.

Magnello, Eileen (2000). *A Century of Measurement: An Illustrated History of The National Physical Laboratory.* Bath: Canopus Publishing.

Manguel, Alberto (1999/1996). *Een geschiedenis van het lezen.* Amsterdam: Ambo. (Originally published in 1996 as *A History of Reading* [London: Harper Collins].)

Mann, Thomas (1936/1913). Death in Venice. In *Stories of Three Decades.* New York: Alfred A. Knopf. (Originally published in 1913 as *Der Tod in Venedig: Novelle* [Berlin: S. Fischer].)

Martin, Andrew (2007). Quiet Please. *Guardian,* January 31, 4–6.

Martin, R. (1991). Geschichte der Schallbewertung. *Zeitschrift für Lärmbekämpfung* 38:151–157.

Martin, W.H. (1929). Decibel—The Name for the Transmission Unit. *Bell System Technical Journal* 8 (January):1–2.

Marwedel, Rainer (1987). *Theodor Lessing: 1872–1933.* Darmstadt: Hermann Luchterhand Verlag.

Marx, Leo (1964). *The Machine in the Garden: Technology and the Pastoral Ideal in America.* London: Oxford University Press.

Massard, Geneviève (1999). Dire le désagrément: Les citadins face à la pollution urbaine, dans la France du XIXe siècle, samenvattingen 4e Internationale Conferentie voor stedengeschiedenis, Venetië, 3–5 september 1998. *Netwerk* 63 (February): 13–14.

Massey, Doreen (2005). *For Space.* London: Sage Publications.

Maur, Karin von (1981). Mondrian und die Musik. In Ulrikre Gauss, ed., *Mondrian: Zeichnungen, Aquarelle, New Yorker Bilder.* Stuttgart: Cantz.

Maur, Karin von (1985). *Vom Klang der Bilder: Die Musik in der Kunst des 20. Jahrhunderts.* Munich: Prestel.

McElligott, Anthony (1999). Walter Ruttmann's 'Berlin: Symphony of a City': Traffic-Mindedness and the City in Interwar Germany. In Malcolm Gee, Tim Kirk, and Jill Steward, eds., *The City in Central Europe: Culture and Society from 1800 to the Present* (pp. 209–238). Aldershot: Ashgate Publishing.

McKennell, A.C. (1963). *Aircraft Noise Annoyance around London (Heathrow) Airport: A survey made in 1961 for the Wilson Committee on the Problem of Noise.* [N.p.]: Central Office of Information.

McKenzie, Dan (1916). *The City of Din: A Tirade against Noise* London: Adlard and Son, Bartholomew Press.

McKenzie, Dan (1928). The Crusade Against Noise. *English Review* 21:691–696.

McLachlan, N.W. (1935). *Noise: A comprehensive Survey from Every Point of View.* Oxford: Oxford University Press/London: Humphrey Milford.

McNeill, William H. (1995). *Keeping Together in Time: Dance and Drill in Human History.* Cambridge, Mass.: Harvard University Press.

McShane, Clay (1979). Transforming the Use of Urban Space: A Look at the Revolution in Street Pavements, 1880–1924. *Journal of Urban History* 5, no. 3:279–307.

McShane, Clay (1994). *Down the Asphalt Path: The Automobile and the American City.* New York: Columbia University Press.

Meer (1935). Meer Stilte: "Anti lawaai"-week in Den Haag. *Nieuwe Rotterdamsche Courant,* September 19.

Meijer Drees, F.J. (1966). De beteugeling van de geluidshinder. *Nederlands Juristenblad* [42], no. 44 (17 December 1966): 1137–1142.

Meister, Franz Joseph (1961). *Geräuschmessungen an Verkehrsflugzeugen und ihre hörpsychologische Bewertung.* Cologne: Westdeutscher Verlag, Arbeitsgemeinschaft für Forschung des Landes Nordrhein-Westfalen, vol. 94.

Melosi, Martin V. (1993). The Urban Environmental Crisis. In Leon Fink, ed., *Major Problems in the Gilded Age and the Progressive Era* (pp. 423–434). Lexington, Mass.: D.C. Heath.

Men (1999). "Men spreke duidelijk . . ." *De Boerhoorn, 3–maandelijks tijdschrift van de historische vereniging "Nörg"* 3, no. 9:24.

Menninghaus, Winfried (1996). Lärm und Schweigen: Religion, moderne Kunst und das Zeitalter des Computers. *Merkur* 50, no. 6:469–479.

Meszaros, Beth (2005). Infernal Sound Cues: Aural Geographies and the Politics of Noise, *Modern Drama* 48, no. 1:118–131.

Met den geluidmeter (1937). Met den geluidmeter op decibeljacht. *De Telegraaf,* October 8, evening edition.

Metz, Gerard (2002). *Hinderlijk geraasch: Geluidbeleid in Breda in de 19e en 20e eeuw.* Groningen: ReGister.

Meyer, H. (No Date). *Achter het Autostuur: Wenken in het belang van rijder en mechanisme.* Amersfoort: Valkenhoff & Co.

Meyer, O.M.T., and Potman, H.P. (1987). *Voor de bestrijding van het lawaai: Een onderzoek naar de vorming van het geluidhinderbeleid in Nederland.* Nijmegen: n.p.

Meyer, Gust de (1997). *Sprekende machines: Geschiedenis van de fonografie en van de muziekindustrie* Leuven: Garant.

Miller, Gale, and Holstein, James A., eds. (1993). *Constructionist Controversies: Issues in Social Problems Theory.* New York: Aldine de Gruyter.

Mills, C.H.G., and Robinson, D.W. (1961). The Subjective Rating of Motor Vehicle Noise. *Engineer,* June 30, 1961, reprint without page-numbers.

Misa, Thomas J., Brey, Philip, and Feenberg, Andrew, eds. (2003). *Modernity and Technology.* Cambridge, Mass.: MIT Press.

Mody, Cyrus C.M. (2001). A Little Dirt Never Hurt Anyone: Knowledge-Making and Contamination in Materials Science. *Social Studies of Science* 31, no. 1 (February): 7–36.

Mody, Cyrus C.M. (2005). The Sounds of Science: Listening to Laboratory Practice. *Science, Technology & Human Values* 30:175–198.

Mom, Gijs (1997). *Geschiedenis van de auto van morgen: Cultuur en techniek van de elektrische auto.* Deventer: Kluwer Bedrijfsinformatie.

Mondriaan, Piet (1917). De jazz en de neo-plastiek. *Internationale Revue* 10, no. 1:421–427.

Mondriaan, Piet (1921a). Le neo-plasticisme. *De Stijl* 4, no. 2:18–23.

Mondriaan, Piet (1921b). De "bruiteurs futuristes italiens" en "het" nieuwe in de muziek. *De Stijl* 4, no. 8:114–118/119, 130–136.

Morgan, Robert P. (1991). *Twentieth Century Music: A History of Musical Style in Modern Europe and America.* New York: W.W. Norton & Co.

Morris, Adalaide, ed. (1997). *Sound States: Innovative Poetics and Acoustical Technologies.* Chapel Hill: University of North Carolina Press.

Morrow, Chester F. (1913). Anti-noise Legislation Now Pending in the City Council and Under Consideration by the Anti-noise Committee. *Bulletin of the Medical and Chirurgical Faculty of Maryland* 5, no. 7:117–121.

Mr Punch (1938). Mr. Punch and Ourselves. *Quiet* 2, no. 9, December, 14.

Mulder, Jan (1997). *Mobieliquette: Vermakelijke verhalen over de etiquette van het mobiel bellen.* Amsterdam: Thomas Rap.

REFERENCES

Mulder, Jan (2000). Zomer. *De Volkskrant,* May 11.

Müller, Peter, and von Schmude, Marcus (2001). Laut, das sind die anderen. *Die Zeit* 33, (August 9): 9–10.

Muziek (1955). Muziek in de fabriek. *De Nederlandse industrie* [4], no. 3:66–67.

Myers, Charles S. (1934). Preface to F.C. Bartlett, *The Problem of Noise* (pp. vii–x). Cambridge: Cambridge University Press.

Nearly, A. (1938). A Nearly Silent Road Drill. *Quiet* 2, no. 9 (December): 15–16.

Needham, Rodney (1967). Percussion and Transition. *Man: A Monthly Record of Anthropological Science* 2:606–614.

Neighbour's, The (1938). The Neighbour's Loud Speaker. *Quiet* 2, no. 8 (April): 31.

Neisius, Erich (1989). *Geschichte der arbeitsmedizinischen Lärmforschung in Deutschland.* Frankfurt am Main: Johann-Wolfgang-Goethe-Universität.

Neuerburg, E.N., and Verfaille, P. (1995). *Schets van het Nederlands milieurecht.* Alphen aan de Rijn: Samson H.D. Tjeenk Willink.

Nieuw (1937). Nieuw wapen tegen den demon van het lawaai. *De Telegraaf,* April 22, Avondblad.

Noise (1935). *Noise Abatement Exhibition.* London: The Anti-Noise League.

Noise (1960). *Noise Abatement Act.* [London: HMSO].

Noise Measurement (1958/1955). *Noise Measurement Techniques.* London: National Physical Laboratory. (Originally published in 1955.)

Noise Units (1975). *Noise Units. Report by a Working Party for the Research Sub-Committee of the Noise Advisory Council.* London: HMSO.

Nuckolls, Janis B. (1999). The Case for Sound Symbolism. *Annual Review of Anthropology* 28:225–252.

Nussbaum, H. Chr. (1912–1913) Geräuschschutz für das Wohnhaus. *Haustechnische Rundschau* 6, no. 23:267–269.

Ommeren, C.W. van, and Hoendervanger, W.J. (1914). *De hinderwet en hare toepassing.* Arnhem: G.W. van der Wiel & Co.

Opdenberg, H. [1938] *Music in Worktime*. [Eindhoven]: Philips.

Overmeer, P. (1999). Lezen in stilte. In F. Messing, P. Overmeer, and H. Willemsen, eds. *Boek en Wijsheid. Beschouwingen over schrijven, lezen en leven* (pp. 183–198). Budel: Uitgeverij Damon.

Pacey, Arnold (1999). *Meaning in Technology*. Cambridge, Mass.: MIT Press.

Parr, Joy (2001). Notes for a More Sensuous History of Twentieth-Century Canada: The Timely, the Tacit, and the Material Body. *Canadian Historical Review* 82, no. 4:720–745.

Passchier-Vermeer, W. (1969). Boekbespreking van "Introduction to the study of noise in industry: International Wrought Copper Council, London, 1968." *Tijdschrift voor Sociale Geneeskunde* 47:282.

Passchier-Vermeer, W. (1972). Een grens tussen veilig en onveilig geluid. *Tijdschrift voor Sociale Geneeskunde* 50, no. 9, supplement 2: 26–32.

Pasternak, Boris (1958). *Doctor Zhivago*. New York: Pantheon.

Payer, Peter (2004). Der Klang von Wien: Zur akustischen Neuordnung des öffentlichen Raumes. *Österreichische Zeitschrift für Geschichtswissenschaften* 15, no. 4:105–131.

Payer, Peter (2006). Vom Geräuch zum Lärm: Zur Geschichte des Hörens im 19. und frühen 20. Jahrhundert. In Volker Bernius, Peter Kemper, and Regina Oehler, eds., *Der Aufstand des Ohrs—die neue Lust am Hören* (pp. 105–119). Göttingen: Vandenhoeck & Ruprecht.

Pepinsky, Abe (1944). The Growing Appreciation of Music and Its Effect upon the Choice of Music in Industry. *Journal of the Acoustical Society of America* 15, no. 3:176–179.

Perzynski, Friedrich (1911). Traute deutsche Hausmusik. *Der Antirüpel* 3, no. 5:22.

Pessers, Dorien (1997). Kakofonie van het moderne leven. *Volkskrant,* December 30.

Peters, Peter (1998). Lopen in Schuberts Winterreise. *Hollands Maandblad* 40, nos. 6–7: 5–13.

Petit, R. (1925). Ballet pour pleyela par George Antheil. *Revue Musicale* 7, no. 3:78–79.

Picker, John M. (2003). *Victorian Soundscapes*. Oxford: Oxford University Press.

Pijper, Willem (1927). Mechanische muziek. *Internationale Revue* 10, no. 1:32–34.

Pinch, Trevor, and Trocco, Frank (2002). *Analog Days: The Invention and Impact of the Moog Synthesizer*. Cambridge, Mass.: Harvard University Press.

Pinkenburg, G. (1903). *Der Lärm in den Städten und seine Verhinderung.* Jena: Verlag von Gustav Fischer.

Politiewijzer (2005). *Politiewijzer 2005–2006.* Maastricht: Politie Limburg Zuid.

Porter, Theodore M. (1995). *Trust in Numbers. The Pursuit of Objecitvity in Science and Public Life.* Princeton, N.J.: Princeton University Press.

Poulussen, Peter (1987). *Van burenlast tot milieuhinder. Het stedelijk leefmilieu, 1500–1800.* Kapellen: DNB/Uitgeverij Pelckmans.

Prause, F.J. (1948). Enige zijden van het vermoeidheidsvraagstuk. *Maandblad voor handelswetenschappen en handelspraktijk* [18], no. 11:145–146.

Preussner, Eberhard (1926). Das sechste Donaueschinger Kammermusikfest. *Die Musik* 18:899–903.

Price, Clair (1934). Hootless London Plans War on Clamor. *New York Times Magazine,* October 21, 8 and 13.

Prins, R.E.J. (1991). Hinderwet en leefklimaat in Kampen, 1875–1940. Master's thesis, Utrecht.

Prioleau, John (1938). The Quieter Motor Car. *Quiet* 2, no. 7 (January): 11–12.

Procedure (1966). *Procedure for Describing Aircraft Noise around an Airport, ISO Recommendation R507.* Geneva: International Organization for Standardization.

Promoting (1937). Promoting Quiet Abroad. *Quiet* 1, no. 6 (October): 30–33.

Promoting (1938a). Promoting Quiet Abroad. *Quiet* 2, no. 7 (January): 34–36.

Promoting (1938b). Promoting Quiet Abroad. *Quiet* 2, no. 8 (April): 33–35.

Purves-Stewart, James (1937). The Educational Influence of the Anti-Noise League. *Quiet* 1, no. 5 (July): 12 and 25.

Pyatt, E.C. (1983). *The National Physical Laboratory—A History.* Bristol: Adam Hilger.

Quiet (1938). Quiet Travel by Rail. *Quiet* 2, no. 9 (December): 17–18.

Randel, Don Michael, ed. (2003). *The Harvard Dictionary of Music.* Cambridge, Mass.: Belknap Press of Harvard University Press.

Rasch, Rudolf, ed. (1987). *Corpus Microtonale. Adriaan Daniël Fokker (1887–1972), selected musical compositions (1948–1972)*. Utrecht: Diapason Press.

Rath, Richard Cullen (2003). *How Early America Sounded*. Ithaca: Cornell University Press.

Ratschläge (1910). Ratschläge für unsere Mittglieder. *Der Antirüpel* 2, no. 6:33.

Razzia (1937). Razzia tegen het lawaai. *De Telegraaf,* October 7, evening edition.

Reducing (1921). Reducing Noise in Factories. *Scientific American,* August 6, 96.

Remarque, Erich Maria (1929). *All Quiet on the Western Front*. Boston: Little, Brown & Co.

Renssen, Henk van (2002). Stiltemuseum. *De Volkskrant,* December 21, 5W.

Report [1914]. *Report of The Society for the Suppression of Unnecessary Noise, 1907–1913*. N.p.: n.p.

Report (1929). *Report of Conference on Road Traffic Noises and Priority of Traffic at Cross Roads*. London: HMSO.

Revill, George (2000). Music and the Politics of Sound: Nationalism, Citizenship, and Auditory space. *Environment and Planning* 18:597–613.

Reynolds, Wynford (1942). *Music While You Work*. London: British Broadcasting Corp.

Rex, Frederick (1935). *Ordinances, Bye-Laws and Regulations of Foreign Cities Providing for the Suppression or Abatement of Noise from Various Causes*. Chicago: Municipal Reference Library.

Richardson, Benjamin W. (1875). Health and Civilization. *Journal of the Society of Arts* 23 (October 15): 948–954.

Richardson, E.G. (1927). *Sound: A Physical Text-Book*. London: Edward Arnold & Co.

Ritsema, Beatrijs (1995). Lawaai. *NRC Handelsblad,* June 14.

Robinson, Anthony (Red.) (1979). *Moderne Ontwikkelingen van de Luchtvaart*. Rotterdam: Lekturama.

Robinson, D.W. (1957). The Subjective Loudness Scale. *Acustica* 7, no. 4:217–233.

Robinson, D.W. (1960). Noise Nuisance: A Quantitative Approach. *Times Science Review* 37 (Autumn): 3–5.

Robinson, D.W. (1964). A Note on the Subjective Evaluation of Noise. *Journal of Sound and Vibration* 1, no. 4:468–473.

Robinson, D.W. (1967). The Subjective Basis for Aircraft Noise Limitation. *Journal of the Royal Aeronautical Society* 71, no. 678 (June): 396–400.

Robinson, D.W. (1973). Rating the Total Noise Environment: Ideal or Pragmatic Approach? *Proceedings of International Congress on Noise as a Public Health Problem, Dubrovnik 1973* (pp. 777–784). N.p.: U.S. Environmental Protection Agency.

Robinson, D.W., Bowsher, J.M., and Copeland, W.C. (1963). On Judging the Noise from Aircraft in Flight. *Acustica* 13, no. 5:324–336.

Robinson, D.W., Copeland, W.C., and Rennie, A.J. (1961). Motor Vehicle Noise Measurement. *Engineer,* March 31, reprint, 2–6.

Rodaway, Paul (1994). *Sensuous Geographies: Body, Sense and Place.* London/New York: Routledge.

Roell, Craigh H. (1989). *The Piano in America 1890–1940.* Chapel Hill: University of North Carolina Press.

Rolland, Romain (1910/1904–1912). *Jean-Christophe,* Volume 2. [Liverpool: John] Holt. (Originally published in 1904–1912 as *Jean-Christophe,* Le Matin [Paris: Librairie Ollendorff].)

Rose, Nikolas (1992). Engineering the Human Soul: Analyzing Psychological Expertise. *Science in Context* 5, no. 2:351–369.

Rosen, George (1974). A Backward Glance at Noise Pollution. *American Journal of Public Health* 64, no. 5:514–517.

Rossum, J.A. van, et al. (1988). *Burenlawaai—een onderzoek naar klachten over burenlawaai.* The Hague: Ministerie van Volkshuisvesting, Ruimtelijke Ordening en Milieubeheer.

Rowland, Stanley (1923). Noise. *The Nineteenth Century and After* 19, no. 559:313–323.

Russolo, Luigi (1986/1916). *The Art of Noises.* New York: Pendragon Press.

Rutters, Herman (1913). De sirene als muziekinstrument. *Caecilia. Maandblad voor Muziek* 70:1–8/A.

Said, Edward W. (1997). From Silence to Sound and Back Again: Music, Literature, and History. *Raritan: A Quarterly Review* 17:1–21.

Saul, Klaus (1996a). "Kein Zeitalter seit Erschaffung der Welt hat so viel und so ungeheuerlichen Lärm gemacht . . .": Lärmquellen, Lärmbekämpfung und Antilärmbewegung

---

im Deutschen Kaiserreich. In Günther Bayerl, Norman Fuchsloch, and Torsten Meyer, eds., *Umweltgeschichte—Methoden, Themen, Potentiale: Tagung des Hamburger Arbeitskreises für Umweltgeschichte, Hamburg 1994* (pp. 187–217). Münster: Waxmann.

Saul, Klaus (1996b). Wider die "Lärmpest": Lärmkritik und Lärmbekämpfung im Deutschen Kaiserreich. In Dittmar Machule, Olaf Mischer, and Arnold Sywottek, eds., *Macht Stadt krank? Vom Umgang mit Gesundheit und Krankheit* (pp. 151–192). Hamburg: Dölling & Galitz.

Schafer, R. Murray (1967). *Ear Cleaning: Notes for an Experimental Music Course.* Toronto: Berandol Music.

Schafer, R. Murray (1969). *The New Soundscape: A Handbook for the Modern Music Teacher.* Toronto: Clark & Cruickshank.

Schafer, R. Murray, ed. (1977a). *European Sound Diary.* Burnaby: Simon Fraser University.

Schafer, R. Murray (1977b). *Five Village Soundscapes.* Burnaby: Simon Fraser University.

Schafer, R. Murray (1993). Soundscape: Design für Ästhetik und Umwelt. In Arnica-Verena Langenmaier et al., eds., *Der Klang der Dinge. Akustik—eine Aufgabe des Design* (pp. 10–27). Munich: Design Zentrum/Verlag Silke Schreiber.

Schafer, R. Murray (1994/1977). *The Soundscape: Our Sonic Environment and the Tuning of the World.* Rochester, Vt.: Destiny Books. (Originally published in 1977 as *The Tuning of the World* [New York: Knopf].)

Schafer, R. Murray (1994). The Soundscape Designer. In Helmi Järviluoma, ed., *Soundscapes: Essays on Vroom and Moo* (pp. 9–18). Tampere: Tampere University.

Schafer, R. Murray (1999). Soundscape, Then and Now. In Henrik Karlsson et al., *From Awareness to Action: Proceedings from "Stockholm, Hey Listen!" Conference on Acoustic Ecology.* Stockholm, June 9–13, 1998 (pp. 25–32). Stockholm: The Royal Swedish Academy of Music.

Schaffer, Simon (1999). Late Victorian Metrology and Its Instrumentation: A Manufactory of Ohms. In M. Biagioli, eds., *The Science Studies Reader* (pp. 457–478). London: Routledge.

Schaffer, Simon (2000). Modernity and Metrology. In Luca Guzzetti, eds., *Science and Power: The Historical Foundations of Research Policies in Europe* (pp. 71–91). Luxembourg: Office for Official Publications of the European Communities.

Schivelbusch, Wolfgang (1979). *The Railway Journey: Trains and Travel in the 19th Century.* Oxford: Basil Blackwell.

Schmidt, Leigh Eric (2004). Sound Christians and Religious Hearing in Enlightenment America. In Mark M. Smith, ed., *Hearing History: A Reader* (pp. 221–246). Athens: University of Georgia Press.

Schmidt, P.H. (1969). Geluidshinderproblemen zoals een keel-neus-oorarts die ziet. *Tijdschrift voor Sociale Geneeskunde* 47, no. 4: 3–6.

Schmidt, Susanne K., and Welre, Raymond (1998). *Coordinating Technology: Studies in the International Standardization of Telecommunications.* Cambridge, Mass.: MIT Press.

Schoemaker, R.L.S. (1929). Het dempen van gevoels- en geluidstrillingen. *Bouwbedrijf* 6, no. 14: 275–276.

Schopenhauer, Arthur (1974/1851). Over lawaai en luidruchtigheid. In *Er is geen vrouw die deugt* (pp. 156–160). Amsterdam: Arbeiderspers. (Originally published as: Ueber Lärm und Geräusch, in *Parerga und Paralipomena, kleine philosophische Schriften* [Berlin: A.W. Hahn].)

Schreuder, Arjen (2006). Slimmer vliegen moet Schiphol redden. *NRC Handelsblad,* January 17, 2.

Schröder, M. (1950). Muziek gedurende het werk. *Mens en Onderneming* [4]:189–195.

Schulz, Theodore John (1972). *Community Noise Ratings.* London: Applied Science Publishers.

Schultze, Ernst (1909). Der Kampf gegen die Lärmplage in den Vereinigten Staaten. *Blätter für Volksgesundheitspflege* 9, no. 8:179–184.

Schünemann, G. (1931). Muziek en techniek. *Maandblad voor Hedendaagsche Muziek* 1:4–6.

Schwartz, Hillel (1998). Beyond Tone and Decibel: The History of Noise. *The Chronicle of Higher Education,* January 9, B8.

Schwartz, Hillel (2003). The Indefensible Ear: A History. In Michael Bull and Les Back, eds., *The Auditory Culture Reader* (pp. 487–501). Oxford: Berg.

Schweighauser, Philipp (2006). *The Noises of American Literature, 1890–1985: Toward a History of Liter/ary Acoustics.* Gainesville: University Press of Florida.

Second (1971). *Second Survey of Aircraft Noise Annoyance around London (Heathrow) Airport.* London: HMSO.

Selvin, Ben (1943). Programming Music for Industry. *Journal of the Acoustical Society of America* 15, no. 2 (October): 131–132.

————

Severini, G. (1919). Eenige denkbeelden over futurisme en cubisme. *De Stijl 2,* no. 3:25–27.

Shapin, Steven (1991). "The Mind Is Its Own Place": Science and Solitude in Seventeenth-Century England. *Science in Context* 4:191–218.

Sherman, Roger W. (1930). Sound Insulation in Apartments. *Architectural Forum* 53:373–378.

Short, A. (1968). A Short History of the Noise Abatement Society. In *International Noise Abatement Congress* (pp. 5–6). London: International Noise Abatement Congress.

Siefert, Marsha (1995). Aesthetics, Technology, and the Capitalization of Culture: How the Talking Machine Became a Musical Instrument. *Science in Context* 8, no. 2:417–449.

Smilor, Raymond Wesley (1971). Cacophony at 34th and 6th: The Noise Problem in America, 1900–1930. *American Studies* 18, no. 1:23–28.

Smilor, Raymond Wesley (1978). *Confronting the Industrial Environment: The Noise Problem in America, 1893–1932.* Dissertation, University of Texas, Austin.

Smilor, Raymond Wesley (1980). Toward an Environmental Perspective: The Anti-Noise Campaign, 1893–1932. In Martin V. Melosi, ed., *Pollution and Reform in American Cities, 1870–1930* (pp. 135–151). Austin: University of Texas Press.

Smith, Bruce R. (1999). *The Acoustic World of Early Modern England. Attending to the O-Factor.* Chicago: University of Chicago Press.

Smith, Mark M. (2001). *Listening to Nineteenth Century America.* Chapel Hill: University of North Carolina Press.

Smith, Mark M. (2003a). Making Sense of Social History *Journal of Social History* 37, no. 1:165–186.

Smith, Mark. M. (2003b). Listening to the Heard Worlds of Antebellum America. In Michael Bull and Les Back, eds., *The Auditory Culture Reader* (pp. 137–163). Oxford: Berg.

Smith, Mark M., ed. (2004). *Hearing History: A Reader.* Athens: University of Georgia Press.

Smith, Meritt Roe, and Leo Marx (1994). *Does Technology Drive History? The Dilemma of Technological Determinism.* Cambridge, Mass.: MIT Press.

Smith, Susan J. (2000). Performing the (Sound) World. *Environment and Planning,* Part D: *Society and Space* 18, no. 5:615–637.

Soft, The (1979). *The Soft Sell: Making Noise a Public Campaign.* Amsterdam: Instituut voor Sociale Kommunikatie.

Some headlines (1962). Some Headlines from Just One Day's Press Cuttings. *Quite Please* 1, no. 3:22–24.

Spitta, Oskar (1941). Ueber den Lärm. *Gesundheits-Ingenieur* 64, no. 2:22–26.

Spotternij (1937). Spotternij of Sotternij. *De Telegraaf,* July 7, 1937.

Spotternij (1938). Spotternij of Sotternij. *De Telegraaf,* April 14, Ochtendblad.

Stallen, Pieter Jan M., and Herman R. van Gunsteren (2002). Schiphol en de illusie van een hinderloze samenleving. In C.T. ten Braven et al., eds., *Essays over luchtvaart: een verkenningsvlucht voor nieuw beleid.* (pp. 166–176). The Hague: Ministerie van Verkeer en Waterstaat.

Starkie D.N.M., and Johnson, D.M. (1975). *The Economic Value of Peace and Quiet.* Lexington, Mass.: Saxon House.

Statistisch Jaarboek (1994). *Statistisch Jaarboek 1994.* Voorburg: Centraal Bureau voor de Statististiek.

Steinhard, Erich (1926). Donaueschingen: Mechanisches Musikfest. *Der Auftakt* 6, no. 8:183–186.

Sterne, Jonathan (2003). *The Audible Past. Cultural Origins of Sound Reproduction.* Durham, N.C.: Duke University Press.

Stevens, S.S. (1956). Calculation of the Loudness of Complex Noises. *The Journal of the Acoustical Society of America* 28, no. 5:807–832.

Stoffers, Manuel (1997). Versnelling, nervositeit en geschiedenis: Over de betekenis van de eigen tijd voor de Duitse cultuurgeschiedschrijving tussen 1890 en 1930. *Kennis en Methode* 21, no. 3:192–203.

Stoller, Paul (1989). *The Taste of Ethnographic Things: The Senses in Anthropology.* Philadelphia: University of Pennsylviania Press.

Storey, David (1961/1960). *Flight into Camden.* New York: MacMillan. (Originally published in 1960 [London: Longmans].)

Stramentov, Constantin (1967). The Architects of Silence. *Unesco Courier* [20], no. 2 (July): 8–12.

Strassburg, D. von (1926). Musikautomate. *Anbruch* 8, no. 2:81–82.

Strauss, H.G. (1937). Noise and Some Legal Remedies. *Quiet* 1, no. 5:17–20.

Strauss, H.G. (1938). The Law and Noise. *Quiet* 2, no. 8 (April):13–14.

Strindberg, August (1967/1886–87). *The Red Room.* London: J.M. Dent & Son. (Originally published in 1886–87 as *Röda rummet,* Part 3 of *Tjensteqvinnans son* [Stockholm: Bonnier].)

Stuckenschmidt, H.H. (1924). Die Mechanisierung der Musik. *Pult und Takstock* 2, no. 1:1–8.

Stuckenschmidt, H.H. (1926a). Mechanische Musik. *Der Auftakt* 6, no. 8:170–173.

Stuckenschmidt, H.H. (1926b). Aeroplansonate (George Antheil). *Der Auftakt* 6, no. 8:178–181.

Stuckenschmidt, H.H. (1926c). Mechanisierung. *Anbruch* 8, no. 8/9:345–346.

Stuckenschmidt, H.H. (1926d). Mechanical Music. *Der Kreis* 3, no. 11:506–508.

Stuckenschmidt, H.H. (1927). Machines—A Vision on the Future. *Modern Music* 4, no. 3:8–14.

Sully, James (1878). Civilisation and Noise. *Fortnightly Review* 24:704–720.

Sunday, The (1960). The Sunday Times, April 8, 1960. *Quiet Please* 1, no.1:38.

Tannen, Deborah, and Saville-Troike, Muriel, eds. (1985). *Perspectives on Silence.* Norwood, N.J.: Ablex Publishing Corp.

Taruskin, Richard (1995). *Text and Act: Essays on Music and Performance.* New York: Oxford University Press.

Taxi's (1937). Taxi's zijn het allerergste. *Het Volk,* November 13, morning edition.

Théberge, Paul (1997). *Any Sound You Can Imagine: Making Music/Consuming Technology.* Hanover, N.H.: Wesleyan University Press.

Thomas, Dylan (1955/1940). *Portrait of the Artist as a Young Dog.* Norfolk, Conn.: New Directions. (Originally published in 1940.)

Thompson, Emily (1995). Machines, Music, and the Quest for Fidelity: Marketing the Edison Phonograph in America, 1877–1925. *Musical Quarterly* 79, no. 1:131–171.

Thompson, Emily (1999). Listening to/for Modernity: Architectural Acoustics and the Development of Modern Spaces in America. In Peter Galison and Emily Thompson, eds., *The Architecture of Science* (pp. 253–280). Cambridge, Mass.: MIT Press.

Thompson, Emily (2002). *The Soundscape of Modernity: Architectural Acoustics and the Culture of Listening in America,1900–1933.* Cambridge, Mass.: MIT Press.

Thompson, R. (1931). American Composers V: George Antheil. *Modern Music* 8, no. 4:17–28.

Thuillier, Guy (1977). *Pour une histoire du quotidien au XIXe siècle en Nivernais.* Paris: La Haye.

Tjaarda Mees, Regnerus (1881). *Staatstoezicht op de fabrieksnijverheid in het belang der openbare rust, veiligheid en gezondheid.* Leiden: Rijksuniversiteit Leiden.

Tjaden, M.E.H. (1929). De kleine electro-motor als vrijbuiter ingetoomd. *Bouwbedrijf* 6, no. 10:190–191.

Toch, Ernst (1926). Musik für mechanische Instrumente. *Anbruch* 8, no. 8/9:346–349.

Toorn, A.J. van der (1949). Over de invloed van muziek bij het werk op werknemer en productie. *Tijdschrift voor Efficiëntie en Documentatie* 19, no. 9:203–207.

Trendelenburg, Ferdinand (1935). *Klänge und Geräusche.* Berlin: Verlag von Julius Springer.

Trobridge, George (1900). The Murder of Sleep. *Westminster Review* [154, no. 3] (September): 298–302.

Truax, Barry, ed. (1978). *The World Soundscape Project's Handbook for Acoustic Ecology.* Vancouver, B.C.: ARC Publications.

Truax, Barry (1984). *Acoustic Communication.* Norwood, N.J.: Ablex Publishing Corporation.

Truax, Barry (1996). Soundscape, Acoustic Communication and Environmental Sound Composition. *Contemporary Music Review* 15, no. 1:49–65.

Uitspraak (1937). Uitspraak inzake den geluidsmeter van prof. Zwikker. *Algemeen Handelsblad,* September 25, evening edition.

Urry, John (1995). *Consuming Places.* London: Routledge.

Urry, John (2000). *Sociology beyond Societies: Mobilities for the Twenty-first Century.* London: Routledge.

Vaillant, Derek (2003). Peddling Noise: Contesting the Civic Soundscape of Chicago, 1980–1913. *Journal of the Illinois State Historical Society* 96, no. 3:257–287.

Valsecchi, Marco (1971). *Landscape Painting in the 19th century.* Greenwich, Conn.: New York Graphic Society.

———

Varèse, Edgar (1917). Que la musique sonne. *Trois cent quatre-vingt-onze* 5:42.

Verberne, A.M.N. (1993). *"Een vergunning onder voorwaarden": Hinderwet en leefklimaat in Tilburg, 1875–1937.* Master's thesis, Tilburg.

Verkeerszondaars (1935). Verkeerszondaars de oorzaak van straatlawaai. De automobilisten tot toeteren gedwongen. *De Telegraaf,* September 19.

Verslag (1934). *Verslag van het 'Anti-lawaai Congres,' georganiseerd te Delft, op 8 november 1934 door de Koninklijke Nederlandsche Automobiel Club in samenwerking met de Geluidstichting.* Delft: KNAC/Geluidstichting.

Verslag (1936). *Verslag van het tweede Anti-Lawaai-Congres te Delft op 21 april 1936 georganiseerd door de Koninklijke Nederlandsche Automobiel Club en de Geluidstichting.* Delft: Geluidstichting.

Vertrauensmänner (1910). Vertrauensmänner der D.A.L.V. *Der Antirüpel* 2 (December): 55–56.

Voorschriften (1966). *Voorschriften en Wenken voor het Ontwerpen van Woningen.* The Hague: Ministerie van Volkshuisvesting en Ruimelijke Ordening.

Vreeken, Rob (2000). Leven. *De Volkskrant,* May 11.

Vuijsje, Herman, and Wouters, Cas (1999). *Macht en gezag in het laatste kwart: Inpakken en wegwezen.* The Hague: Sociaal en Cultureel Planbureau.

Wagner, Karl Willy (1931). *Geräusch und Lärm: Mitteilung aus dem Heinrich-Hertz-Institut für Schwingungsforschung.* Berlin: Verlag der Akademie der Wissenschaften in Kommission bei Walter de Gruyter U. Co.

Wagner, Karl Willy (1935). Fortschritte in der Geräuschforschung und Lärmabwehr. *Zeitschrift des Vereines Deutscher Ingenieure* 79, no. 18 (May 4): 531–540.

Wagner, Karl Willy (1936a). Praktische Wege der Lärmabwehr. *Forschungen und Fortschritte* 12, no. 3 (January): 41–43.

Wagner, Karl Willy (1936b). Grundlagen der Lärmabwehr. *Forschungen und Fortschritte* 12, no. 1 (January): 12–13.

Wagner, Karl Willy (1938). Vorschlag zu einer praktischen Definition der Lautheit. *Hochfrequenztechnik und Elektroakustik* 45:14–18.

Warwick, Alan R. (1968). *A Noise of Music.* London: Queen Anne Press.

Watkins, Glenn (1988). *Soundings: Music in the Twentieth Century.* New York: Macmillan.

Watson-Verran, H., and Turnbull, D. (1995). Science and Other Indigenous Knowledge Systems. In Sheila Jasanoff et al., *Handbook of Science and Technology Studies* (pp. 115–139). London: Sage.

Weidenaar, Reynold (1995). *Magic Music from the Telharmonium.* Metuchen, N.J.: Scarecrow Press.

Weil, I. (1927). The Noise-Makers. *Modern Music* 5, no. 2:24–28.

Weissmann, A. (1927). Mensch und Machine. *Die Musik* 20:103–107.

Weissmann, A. (1928). *Die Entgötterung der Musik.* Stuttgart: Deutsche Verlags-Anstalt.

Wendt, H. (1910). Die Not der Zeit. *Die Gartenlaube* 58:122–123.

White, Hayden (1990). *The Content of the Form: Narrative Discourse and Historical Representation.* Baltimore: Johns Hopkins University Press.

Whiteclay Chambers II, John (1992). *The Tyranny of Change: America in the Progressive Era, 1890–1920.* New York: St. Martin's Press.

Whitesitt, Linda (1983). *The Life and Music of George Antheil, 1900–1959.* Ann Arbor: UMI Research.

Wiethaup, Hans (1966). Lärmbekämpfung in historischer Sicht: Vorgeschichtliche Zeit—Zeitalter der alten Kulturen usw. *Zentralblatt für Arbeitsmedizin und Arbeitsschutz* 16, no. 5:120–124.

Wiethaup, Hans (1967). *Lärmbekämpfung in der Bundesrepublik Deutschland.* Cologne: Carl Heymanns Verlag

Wigge, Heinrich (1936). *Lärm: Die Grundtatsachen der Schalltechnik. Lärmstörungen—Lärmschütz.* Leipzig: Dr. Max Jänecke, Verlagsbuchhandlung.

Will (1939). Will Penalties Stop Noise? *Quiet* 2, no. 10:17.

Willis, Paul (1980). Shop Floor Culture, Masculinity and the Wage Form. In John Clarke, Charles Critcher, and Richard Johnson, eds., *Working-Class Culture: Studies in History and Theory* (pp. 185–198). New York: St. Martin's Press.

Wilson, A.H. (1963). *Noise: Final Report. Presented to Parliament by the Lord President of the Council and Minister for Science by Command of Her Majesty, July 1963.* London: HMSO.

Wit, Onno de, and Bruhèze, Adri A.A. de la (2002). Bedrijfsmatige bemiddeling: Philips en Unilever en de marketing van radio's, televisies en snacks in Nederland in de twintigste eeuw. *Tijdschrift voor Sociale Geschiedenis* 28, no. 3:347–372.

Woolf, Daniel R. (2004). Hearing Renaissance England. In Mark M. Smith, ed. (2004). *Hearing History: A Reader* (pp. 112–135). Athens: University of Georgia Press.

Woolf, Virginia (1925). *Mrs. Dalloway.* New York: Harcourt, Brace and World.

World's Plague, The (1928). The World's Plague of Noise. *Literary Digest,* October 6, 18–19.

Wyatt, S., and Langdon, J.N. (1937). *Fatigue and Boredom in Repetitive Work.* London: HMSO.

Wynne, Shirley W. (1930). New York City's Noise Abatement Commission. *Journal of the Acoustical Society of America* 2, no. 1:12–17.

Y.Y. (1935). Less Noise, Please. *New Statesman and Nation,* August 31, 274–275.

Zachte (1938). "Zachte" hoorn luider dan de "harde." *De Telegraaf,* February 11.

Zeller, Werner (1936). Anti-Noise Work in Berlin. *Quiet* 1, no. 2 (June): 10–13.

Zeller, Werner (1950). *Technische Lärmabwehr.* Stuttgart: Alfred Kröner Verlag.

Zwikker, C. (1936). *Acoustische problemen bij de betonbouw: Voordracht, gehouden voor de Beton-vereeniging op 4 maart 1936 te 's-Gravenhage.* Delft: Geluidstichting.

Zwikker, C. (1938). De Geluidstichting. *In de Driehoek: Bijdragen op het gebied van de Volkshuis-vesting in Nederland* 3, no. 6:248–253.

Zwikker, C. (1939/40). *Geluidmetingen.* Publicatie No. 24. Delft: Geluidstichting.

Zwikker, C. (1972). 50 jaar akoestiek. In *Najaarsvergadering 1971, Publikatie nr. 22* (pp. 74–100). Delft: Nederlands Akoestisch Genootschap.

INSIDE TECHNOLOGY
edited by Wiebe E. Bijker, W. Bernard Carlson, and Trevor Pinch

Herbert Gottweis, *Governing Molecules: The Discursive Politics of Genetic Engineering in Europe and the United States*

Joshua M. Greenberg, *From Betamax to Blockbuster: Video Stores and the Invention of Movies on Video*

Kristen Haring, *Ham Radio's Technical Culture*

Gabrielle Hecht, *The Radiance of France: Nuclear Power and National Identity after World War II*

Kathryn Henderson, *On Line and On Paper: Visual Representations, Visual Culture, and Computer Graphics in Design Engineering*

Christopher R. Henke, *Cultivating Science, Harvesting Power: Science and Industrial Agriculture in California*

Christine Hine, *Systematics as Cyberscience: Computers, Change, and Continuity in Science*

Anique Hommels, *Unbuilding Cities: Obduracy in Urban Sociotechnical Change*

David Kaiser, editor, *Pedagogy and the Practice of Science: Historical and Contemporary Perspectives*

Peter Keating and Alberto Cambrosio, *Biomedical Platforms: Reproducing the Normal and the Pathological in Late-Twentirth-Century Medicine*

Eda Kranakis, *Constructing a Bridge: An Exploration of Engineering Culture, Design, and Research in Nineteenth-Century France and America*

Christophe Lécuyer, *Making Silicon Valley: Innovation and the Growth of High Tech, 1930–1970*

Pamela E. Mack, *Viewing the Earth: The Social Construction of the Landsat Satellite System*

Donald MacKenzie, *Inventing Accuracy: A Historical Sociology of Nuclear Missile Guidance*

Donald MacKenzie, *Knowing Machines: Essays on Technical Change*

Donald MacKenzie, *Mechanizing Proof: Computing, Rick, and Trust*

Donald MacKenzie, *An Engine, Not a Camera: How Financial Models Shape Markets*

Maggie Mort, *Building the Trident Network: A Study of the Enrollment of People, Knowledge, and Machines*

Peter D. Norton, *Fighting Traffic: The Dawn of the Moter Age in the American City*

Nelly Oudshoorn and Trevor Pinch, editors, *How Users Matter: The Co-Construction of Users and Technology*

Shobita Parthasarathy, *Building Genetic Medicine: Breast Cancer, Technology, and the Comparative Politics of Health Care*

Paul Rosen, *Framing Production: Technology, Culture, and Change in the British Bicycle Industry*

Susanne K. Schmidt and Raymund Werle, *Coordinating Technology: Studies in the International Standardization of Telecommunications*

Wersley Shrum, Joel Genuth, and Ivan Chompalov, *Structures of Scientific Collaboration*

Charis Thompson, *Making Parents: The Ontological Choreography of Reproductive Technology*

Dominique Vinck, editor, *Everyday Engineering: An Ethnography of Design and Innovation*

INDEX

Kaye, George William Clarkson, 240, 242
  aircraft and, 203–204
  traffic and, 106, 108, 121
  zoning and, 246
Kennedy, Foster, 115
Kerse, C. S., 57, 59, 69, 73, 108, 214, 242
Kettlemusic, 33
Kingsbury, B. A., 106
Klaxon horns, 118, 204
Kleinhoonte van Os, G. J., 219, 221,
  224–225
Koelma, M., 163
Kosten, C. W., 182–183, 230
  Advisory Committee on Aircraft Noise
    Nuisance and, 218–220
  aircraft noise and, 218–228
  airport planning and, 219–220
  background of, 218
  comprehensive approach of, 219
  decibel scale and, 221–222, 227–228
  international approach of, 218
  International Civil Aviation Organization
    (ICAO) and, 220–222
  International Standardization Organization
    (ISO) and, 218–223, 227–228
  PNdB (Perceived Noise level in decibels)
    and, 221–223, 227–228
  runway reports and, 219
Kostenunit (noise index), 193–194
  aircraft and, 224–228
  step-by-step numbers and, 229–230
Kowar, H., 139
Krenek, Ernst, 154–155

Krömer, S., 72
Kruithof, C. L., 33
Ku, J. H., 194
Kuhn, Thomas, 207
Kuiper, J. P., 73
Kurtze, G., 76

Laan, M., van der, 5
Labor
  Antinoise leagues and, 102–103
  city noise and, 93–104, 133–134
  German Workmen's Compensation Act
    and, 73
  gramophones and, 168
  hearing loss and, 69–81
  industrial hygienists and, 74–75, 79–80
  laboratory employees and, 78
  Laird study and, 113–115
  listening to machinery and, 76–78
  Morgan study and, 113–115
  music and, 81–87
  night shifts and, 54
  nuisance legislation and, 55–57, 64–68
  productivity and, 82–87
  putative sound effects and, 172–173
  radio and, 168–169
  sleep deprivation and, 65–68
  textile trade and, 113–114
Labor Vitamins (music program), 87
Labrie, A., 37
Laird, Donald A., 113–115
Lammers, B., 76, 79
Lamprecht, Karl, 101

Science, 187. *See also* Technology
blood alcohol measurement and, 160
British aircraft noise standard and, 203–218
culture of control and, 160
Dutch aircraft noise standard and, 218–228
noise meters and, 175–180
OECD Directorate for Scientific Affairs
and, 220
pathologization of complaints and, 173
physiological/psychological studies and,
171–178
psychoacoustics and, 173–178
science and technology studies (STS) and,
256–259
social surveys and, 203–218
traffic monitoring and, 160
Wilson Report and, 190
Science and Technology Studies (STS), 20,
256–259
Self-diffusion, 96–97
Self-realization, 250–251
Selvin, B., 86
Senses, 2, 12, 21, 24, 39, 97, 143, 161, 256,
259.
Severini, Gino, 142
Sexuality, 14
Shakespearean theater, 42
Shapin, Steven, 30, 62
Sheet Music of Dawn, The (journal), 151
Shouting, 32–33
Siefert, M., 26
Silence, 5, 135, 261
air-borne vs. structure-borne, 174
campaigns for, 2, 124–133, 175, 177–179

class and, 238–239
cultural connotations and, 31–41
fear of, 2, 10–11
hearing and, 75
hospitals and, 61–66, 88–89, 224
intellectuals and, 94, 96
islands of, 4, 55–69, 218, 246, 250, 255–256
order and, 38
out of place, 240
pre-modern Europe and, 38
psychoacoustics and, 173–178
public campaigns for, 124–133
public education and, 92–93
religion and, 38–39, 62
scholars and, 61–62
schools and, 60–63, 88–89, 224
science and, 39
self-controlled speech and, 38–39
technology and, 39
truth and, 61–62
turbojet engines and, 201–202
visual regime and, 39–40
windmill symbolism and, 27–30
Silence Brigade, 177, 179
Silenta, 175, 177–179, 188–189
Simon Fraser University, 5, 265n10
Singing of the Train, The (Berlioz), 139
Slavery, 33–34, 38
Sleep, 163
aircraft and, 198
importance of, 172
modernization and, 53–57
neighbor noise and, 167–168
tradesmen and, 65–68

neighbor noise and, 164, 172

traffic and, 104, 108

*Village Bells: Sound and Meaning in the
Nineteenth-century French Countryside*
(Corbin), 13–14

Visual regime, 2, 11–12, 39–40, 239–240

Voorschriften, 182

Vreeken, Rob, 5

Vuijsje, Herman, 250–251

Wagner, Karl Willy, 82, 105, 108–109, 131

War, 32–34

Water pollution, 242

Watkins, G., 141

Watson-Verran, H., 80

Weidenaar, R., 139

Weil, Irving, 156

Weissmann, Adolf, 155–156

Welre, R., 203, 229

Welte-Mignons, 154–156

Wendt, H., 96

Western culture, 3

acceptable noise and, 235–240

city noise, 93–104 (*see also* City noise)

class and, 31, 40, 92, 136, 154, 162, 165–
169, 238–239, 256

complaining in style and, 259–260

connotations of noise and silence, 31–41

constructionism and, 18

control and, 160

fear of silence and, 2, 10–11

individualization and, 240–242, 250–255

linguistics and, 31–32

nuisance legislation and, 24–25, 55–69 (*see
also* Nuisance legislation)

olympian perspective on, 234

print and, 11–12

putative difficulties and, 9–13

Reformation and, 38–39

Renaissance and, 62

science and technology studies (STS) and,
256–259

self-controlled speech and, 38–39

self-realization and, 250–251

silence campaigns and, 124–133

stench and, 13–17

studying public problems and, 256–259

technological determinism and, 20

utopian perspective and, 234

visual regime of, 2, 11–12, 39–40, 239–
240

windmill symbolism and, 27–30

Western Electric Holland, 174

Westminster Review, 53

Whistlers, 142

White, Hayden, 234

Whiteclay Chambers II, John, 103

Whitesitt, L., 148

Wien scale, 108

Wiethaup, Hans, 56, 69, 73, 187

Willis, P., 79

Wilson, Alan

aircraft and, 197, 202–203, 208–214, 219

industry and, 73

Wilson Report on Noise, 190

Windmill symbolism, 27–30

Wit, O. de, 168

Witherspoon, 34–35

Woolf, Virginia, 39, 45, 48

World Forum for Acoustic Ecology, 22

Printed in the United States
By Bookmasters